AF560492

Environmental Nanotechnology: Applications and Impacts

Environmental Nanotechnology: Applications and Impacts

Dr. D.K. Chauhan

RANDOM PUBLICATIONS
NEW DELHI (INDIA)

Environmental Nanotechnology: Applications and Impacts

978-93-5111-569-4

Published in 2015 in India by

Reprint 2019

RANDOM PUBLICATIONS

4376-A/4B, Gali Murari Lal, Ansari Road
New Delhi-110 002
Phone : +9111-43580356, 011-23289044, 011-43142548
e-mail: sales@randompublications.com,
info@randompublications.com, randomexports@gmail.com

Type Setting by : Friends Media, Delhi-110089
Printed at : Mehra Printers, Delhi-110 092

Preface

Nanotechnology's environmental impact can be split into two aspects: the potential for nanotechnological innovations to help improve the environment, and the possibly novel type of pollution that nanotechnological materials might cause if released into the environment. The environmental impact of nanotechnology is the possible effects that the use of nanotechnological materials and devices will have on the environment. As nanotechnology is an emerging field, there is great debate regarding to what extent industrial and commercial use of nanomaterials will affect organisms and ecosystems.

Nanotechnological products, processes and applications are expected to contribute significantly to environmental and climate protection by saving raw materials, energy and water as well as by reducing greenhouse gases and hazardous wastes. Using nanomaterials therefore promises certain environmental benefits and sustainability effects. However, that nanotechnology currently plays a rather subordinate role in environmental protection, whether it be in research or in practical applications. Environmental engineering companies themselves attach only limited importance to nanotechnology in their respective fields.

Nanopollution is a generic name for waste generated by nanodevices or during the nanomaterials manufacturing process. Ecotoxicological impacts of nanoparticles and the potential for bioaccumulation in plants and microorganisms is a subject of current research, as nanoparticles are considered to present novel environmental impacts. Of the US$710 million spent in 2002 by the U.S. government on nanotechnology research, $500,000 was spent on environmental impact assessments.

Molecular nanotechnology is a speculative subfield of nanotechnology regarding the possibility of engineering molecular assemblers, machines which could re-order matter at a molecular or atomic scale. Regarding the risks from molecular manufacturing, an often cited worst-case scenario is "grey goo", a hypothetical substance into which the surface of the earth might be transformed by self-replicating nanobots running amok.

This book is an authoritative, in-depth exploration of the environmental consequences of nano technology.

I would like to thank my team for standing beside me throughout my career and writing this book. My special thanks go to "Random Publications" who have published the book.

– Dr. D.K. Chauhan

Contents

1

Environmentl al Impact of Nanotechnology

The environmental impact of nanotechnology is the possible effects that the use of nanotechnological materials and devices will have on the environment. As nanotechnology is an emerging field, there is great debate regarding to what extent industrial and commercial use of nanomaterials will affect organisms and ecosystems.

Nanotechnology's environmental impact can be split into two aspects: the potential for nanotechnological innovations to help improve the environment, and the possibly novel type of pollution that nanotechnological materials might cause if released into the environment.

NANOPOLLUTION

Nanopollution is a generic name for waste generated by nanodevices or during the nanomaterials manufacturing process. Ecotoxicological impacts of nanoparticles and the potential for bioaccumulation in plants and microorganisms is a subject of current research, as nanoparticles are considered to present novel environmental impacts. Of the US$710 million spent in 2002 by the U.S. government on nanotechnology research, $500,000 was spent on environmental impact assessments.

The capacity for nanoparticles to function as a transport mechanism also raises concern about the transport of heavy metals and other environmental contaminants. Two areas of concern can be identified. First, in their free form nanoparticles can be released into the air or water during production, or production accidents, or as waste by-product of production, and ultimately accumulate in the soil, water, or plant life. Second, in fixed form, where they are part of a manufactured substance or product, they will ultimately have to be recycled or disposed of as waste.

Scrinis raises concerns about nano-pollution, and argues that it is not currently possible to "precisely predict or control the ecological impacts of the release of these nano-products into the environment." A May 2007 Report to the UK Department for Environment, Food and Rural Affairs noted concerns about the toxicological impacts of nanoparticles in relation to both hazard and exposure. The report recommended comprehensive toxicological testing and independent performance tests of fuel additives. Risks have been identified by Uskokovic in 2007. Concerns have also been raised about Silver Nano technology used by Samsung in a range of appliances such as washing machines and air purifiers.

LIFE CYCLE RESPONSIBILITY

To properly assess the health hazards of engineered nanoparticles the whole life cycle of these particles needs to be evaluated, including their fabrication, storage and distribution, application and potential abuse, and disposal. The impact on humans or the environment may vary at different stages of the life cycle.

The Royal Society report identified a risk of nanoparticles or nanotubes being released during disposal, destruction and recycling, and recommended that "manufacturers of products that fall under extended producer responsibility regimes such as end-of-life regulations publish procedures outlining how these materials will be managed to minimize possible human and environmental exposure" (p.xiii). Reflecting the challenges for ensuring responsible life cycle regulation, the Institute for Food and Agricultural Standardshas proposed standards for nanotechnology research and development should be integrated across consumer, worker and environmental standards. They also propose thatNGOs and other citizen groups play a meaningful role in the development of these standards.

ENVIRONMENTAL APPLICATIONS OF NANOTECHNOLOGY

ENVIRONMENTAL REMEDIATION

Nanoremeditation is the use of nanoparticles for environmental remediation. Nanoremediation has been most widely used for groundwater treatment, with additional extensive research in wastewater treatment. Nanoremediation has also been tested for soil and sediment cleanup. Even more preliminary research is exploring the use of nanoparticles to remove toxic materials from gases.

Some nanoremediation methods, particularly the use of nano zero-valent iron for groundwater cleanup, have been deployed at full-scale cleanup sites. Nanoremediation is an emerging industry; by 2009, nanoremediation technologies had been documented in at least 44 cleanup sites around the world, predominantly in the United States.During nanoremediation, a nanoparticle agent must be brought into contact with the target contaminant under conditions that allow a detoxifying or immobilizing reaction. This process typically involves a pump-and-treat process or *in situ* application. Other methods remain in research phases.

WATER FILTRATION

Nanofiltration is a relatively recent membrane filtration process used most often with low total dissolved solids water such as surface water and fresh groundwater, with the purpose of softening (polyvalent cation removal) and removal of disinfection by-product precursors such as natural organic matter and synthetic organic matter. Nanofiltration is also becoming more widely used in food processing applications such as dairy, for simultaneous concentration and partial (monovalent ion) demineralisation.

Nanofiltration is a membrane filtration based method that uses nanometer sized cylindrical through-pores that pass through the membrane at a 90°. Nanofiltration membranes have pore sizes from 1-10 Angstrom, smaller than that used in microfiltration and ultrafiltration, but just larger than that in reverse osmosis. Membranes used are predominantly created from polymer thin films. Materials that are commonly used include polyethylene terephthalate or metals such as aluminum. Pore dimensions are controlled by pH, temperature and time during development with pore densities ranging from 1 to 106 pores per cm^2. Membranes made from polyethylene terephthalate and other similar materials, are referred to as "track-etch" membranes, named after the way the pores on the membranes are made. "Tracking" involves bombarding the polymer thin film with high energy particles. This results in making tracks that are chemically developed into the membrane, or "etched" into the membrane, which are the pores. Membranes created from metal such as alumina membranes, are made by electrochemically growing a thin layer of aluminum oxide from aluminum metal in an acidic medium.

Some water-treatment devices incorporating nanotechnology are already on the market, with more in development. Low-cost nanostructured separation membranes methods have been shown to be effective in producing potable water in a recent study.

ENERGY

Research is underway to use nanomaterials for purposes including more efficient solar cells, practical fuel cells, and environmentally friendly batteries. The most advanced nanotechnology projects related to energy are: storage, conversion, manufacturing improvements by reducing materials and process rates, energy saving (by better thermal insulation for example), and enhanced renewable energy sources.

Research is ongoing to use nanowires and other nanostructured materials with the hope of to create cheaper and more efficient solar cells than are possible with conventional planar silicon solar cells. Another example is the use of fuel cells powered by hydrogen, potentially using a catalyst consisting of carbon supported noble metal particles with diameters of 1-5 nm. Materials with small nanosized pores may be suitable for hydrogen storage. Nanotechnology may also find applications in batteries, where the use ofnanomaterials may enable batteries with higher energy content or supercapacitors with a higher rate of recharging.

THE APPLICATION OF NANOTECHNOLOGY TO ENVIRONMENTAL ISSUES

In trying to help our ailing environment, nanotechnology researchers and developers are pursuing the following avenues:

Generating less pollution during the manufacture of materials. One example of this is how researchers have demonstrated that the use of silver nanoclusters as catalysts can significantly reduce the polluting byproducts generated in the process used to manufacture propylene oxide. Propylene oxide is used to produce common materials such as plastics, paint, detergents and brake fluid.

Producing solar cells that generate electricity at a competitive cost. Researcher have demonstrated that an array of silicon nanowires embedded in a polymer results in low cost but high efficiency solar cells. This, or other efforts using nanotechnology to improve solar cells, may result in solar cells that generate electricity as cost effectively as coal or oil.

Increasing the electricity generated by windmills. Epoxy containing carbon nanotubes is being used to make windmill blades. The resulting blades are stronger and lower weight and therefore the amount of electricity generated by each windmill is greater.

Cleaning up organic chemicals polluting groundwater. Researchers have shown that iron nanoparticles can be effective in cleaning up organic solvents that are polluting groundwater. The iron nanoparticles disperse throughout the body of water and decompose the organic solvent in place. This method can be more effective and cost significantly less than treatment methods that require the water to be pumped out of the ground.

Cleaning up oil spills. Using photocatalytic copper tungsten oxide nanoparticles to break down oil into biodegradable compounds. The nanoparticles are in a grid that provides high surface area for the reaction, is activated by sunlight and can work in water, making them useful for cleaning up oil spills.

Capturing carbon dioxide in power plant exhaust. Researchers are developing nanostructred membranes designed to capture carbon dioxide in the exhaust stacks of power plants instead of releasing it into the air.

Clearing volatile organic compounds (VOCs) from air. Researchers have demonstrated a catalyst that breaks down VOCs at room temperature. The catalyst is composed of porous manganese oxide in which gold nanoparticles have been embedded.

Reducing the cost of fuel cells. Changing the spacing of platinum atoms used in a fuel cell increases the catalytic ability of the platinum. This allows the fuel cell to function with about 80 per cent less platinum, significantly reducing the cost of the fuel cell.

Storing hydrogen for fuel cell powered cars. Using graphene layers to increase the binding energy of hydrogen to the graphene surface in a fuel tank results in a higher amount of hydrogen storage and a lighter weight fuel tank. This could help in the development of practical hydrogen-fueled cars.

NANOTECHNOLOGY AND THE ENVIRONMENT

Nanotechnological products, processes and applications are expected to contribute significantly to environmental and climate protection by saving raw materials, energy and water as well as by reducing greenhouse gases and hazardous wastes. Using nanomaterials therefore promises certain environmental benefits and sustainability effects. Note, however, that nanotechnology currently plays a rather subordinate role in environmental protection, whether it be in research or in practical applications. Environmental engineering companies themselves attach only limited importance to nanotechnology in their respective fields.

POTENTIAL ENVIRONMENTAL BENEFITS

Rising prices for raw materials and energy, coupled with the increasing environmental awareness of consumers, are responsible for a flood of products on the market that promise certain advantages for environmental and climate protection. Nanomaterials exhibit special physical and chemical properties that make them interesting for novel, environmentally friendly products.

Examples include the increased durability of materials against mechanical stress or weathering, helping to increase the useful life of a product; nanotechnology-based dirt- and water-resistant coatings to reduce cleaning efforts; novel insulation materials to improve the energy efficiency of buildings; adding nanoparticles to a material to reduce weight and save energy during transport.

In the chemical industry sector, nanomaterials are applied based on their special catalytic properties in order to boost energy and resource efficiency, and nanomaterials can replace environmentally problematic chemicals in certain fields of application.

High hopes are being placed in nano-technologically optimized products and processes for energy production and storage; these are currently in the development phase and are slated to contribute significantly to climate protection and solving our energy problems in the future.

In most commercially available "nano-consumer products", environmental protection is not the primary goal. Neither textiles with nanosilver to combat perspiration odour, nor especially stable golf clubs with carbon nanotubes, help protect the environment. Manufacturers often promise such advantages, typically without providing the relevant evidence. Examples include self-cleaning surface coatings or textiles with spot protection, with are advertized as reducing the cleaning effort and therefore saving energy, water and cleaning agents.

Emphasis is often placed on the sustainable potential of nanotechnology. Nonetheless, this usually reflects unsubstantiated expectations. Determining the actual effects of a product on the environment – both positive and negative – requires examining the entire life cycle from production of the raw material to disposal at the end of the life cycle. As a rule, the descriptions of environmental benefits fail to consider the amount of resources and energy consumed in producing the products. (read more: "Nanotechnology and the environment - Potential benefits and sustainability effects")

SPECIFIC EXAMPLES OF NANOTECHNOLOGY APPLICATIONS THAT BENEFIT THE ENVIRONMENT

Nanotechnology could make battery recycling economically attractive

Many batteries still contain heavy metals such as mercury, lead, cadmium, and nickel, which can contaminate the environment and pose a potential threat to human health when batteries are improperly disposed of. Not only do the billions upon billions of batteries in landfills pose an environmental problem, they also are a complete waste of a potential and cheap raw material. Researchers have managed to recover pure zinc oxide nanoparticles from spent Zn-MnO_2 batteries alkaline batteries.

Nanomaterials for radioactive waste clean-up in water

Scientists are working on nanotechnology solution for radioactive waste cleanup, specifically the use of titanate nanofibres as absorbents for the removal of radioactive ions from water. Researchers have also reported that the unique structural properties of titanate nanotubes and nanofibres make them superior materials for removal of radioactive cesium and iodine ions in water.

Nanotechnology-based solutions for oil spills

Conventional clean-up techniques are not adequate to solve the problem of massive oil spills. In recent years, nanotechnology has emerged as a potential source of novel solutions to many of the world's outstanding problems. Although the application of nanotechnology for oil spill cleanup is still in its nascent stage, it offers great promise for the future. In the last couple of years, there has been particularly growing interest worldwide in exploring ways of finding suitable solutions to clean up oil spills through use of nanomaterials.

Water applications

The potential impact areas for nanotechnology in water applications are divided into three categories – treatment and remediation, sensing and detection, and pollution prevention –and the improvement of desalination technologies is one key area thereof. Nanotechnology-based water purification devices have the potential to transform the field of desalination, for instance by using the*ion concentration polarization* phenomenon.

Another, relatively new method of purifying brackish water is *capacitive deionization* (CDI) technology. The advantages of CDI are that it has no secondary pollution, is cost-effective and energy efficient. Nanotechnology researchers have developed a CDI application that uses graphene-like nanoflakes as electrodes for capacitive deionization. They found that the graphene electrodes resulted in a better CDI performance than the conventionally used activated carbon materials.

Carbon dioxide capture

Before CO_2 can be stored in Carbon dioxide Capture and Storage (CCS) schemes, it must be separated from the other waste gases resulting from combustion or industrial processes. Most current methods used for this type of filtration are expensive and require the use of chemicals. Nanotechnology techniques to fabricate nanoscale thin membranes could lead to new membrane technology that could change that.

Hydrogen production from sunlight - artificial photosynthesis

Companies developing hydrogen-powered technologies like to wrap themselves in the green glow of environmentally friendly technology that will save the planet. While hydrogen fuel indeed is a clean energy carrier, the source of that hydrogen often is as dirty as it gets. The problem is that you can't dig a well to tap hydrogen, but hydrogen has to be produced, and that can be done using a variety of resources.

The dirtiest method – at least until highly efficient carbon capture and sequestration technologies are developed – is the gasification of coal (read more: "Nanotechnology could clean up the hydrogen car's dirty little secret"). The cleanest by far would be renewable energy electrolysis: using renewable energy technologies such as wind, solar, geo- and hydrothermal power to split water into hydrogen and oxygen.

Artificial photosynthesis, using solar energy to split water generating hydrogen and oxygen, can offer a clean and portable source of energy supply as durable as the sunlight. It takes about 2.5 volts to break a single water molecule down into oxygen along with negatively charged electrons and positively charged protons. It is the extraction and separation of these oppositely charged electrons and protons from water molecules that provides the electric power.

Working on the nanoscale, researchers have shown that an inexpensive and environmentally benign inorganic light harvesting nanocrystal array can be combined with a low-cost electrocatalyst that contains abundant elements to fabricate an inexpensive and stable system for photoelectrochemical hydrogen production.

GREEN NANOTECHNOLOGY

There is a general perception that nanotechnologies will have a significant impact on developing 'green' and 'clean' technologies with considerable environmental benefits. The best examples are the use of nanotechnology in areas ranging from water treatment to energy breakthroughs and hydrogen applications. As a matter of fact, renewable energy applications probably are the areas where nanotechnology will make its first large-scale commercial breakthroughs (see: Nanotechnology applications could provide the required energy breakthroughs).

Conflicting with this positive message is the growing body of research that raises questions about the potentially negative effects of engineered nanoparticles on human health and the environment. This area includes the actual processes of manufacturing nanomaterials and the environmental footprint they create, in absolute terms and in comparison with existing industrial manufacturing processes (read more: "Nanotechnology - not that green?"). In order to make any conclusive observations, 'green' nanotechnology requires a full life cycle assessment like any other industrially manufactured products.

A white paper ("Green Nanotechnology Challenges And Opportunities") issued by the ACS Green Chemistry Institute in partnership with the Oregon Nanoscience and Microtechnologies Institute addresses the critical challenges to advancing greener nanotechnology.

Researchers agree that the safest possible future for advancing nanotechnology in a sustainable world can be reached by using green chemistry. Green chemistry means designing chemical products and processes in a way that reduces or eliminates hazardous substances from the beginning to end of a chemical product's life cycle. The practice began in the United States with the passage of the Pollution Prevention Act of 1990, which established a national policy to prevent or reduce pollution at its source whenever feasible. Reducing pollution at the source, according to the act, "is fundamentally different and more desirable" than managing waste and controlling pollution. Since then, the EPA Green Chemistry Programme has built collaborations with academia, industry, other government agencies, non-governmental organizations and international partners to promote pollution prevention through green chemistry.

As the report "Green Nanotechnology: It's easier than you think" (pdf) states: "Green nanotechnology offers the opportunity to head off adverse effects before they occur. Green nanotechnology can proactively influence the design of nanomaterials and products by eliminating or minimizing pollution from the production of the nanomaterials, taking a life cycle approach to nanoproducts to estimate and mitigate where environmental impacts might occur in the product chain, designing toxicity out of nanomaterials and using nanomaterials to treat or remediate existing environmental problems. Green nanotechnology does not arise de novo; rather, it builds on the principles of green chemistry and green engineering and focuses them through a new lens on the unique and often counterintuitive effects that occur in nanoscale materials.

Apart from the obvious areas of using nanomaterials in the areas of solar cells, biofuels and fuel cells, green nanotechnology applications might involve a clean production process, such as synthesizing nanoparticles with sunlight or the recycling of industrial waste products into nanomaterials, such as turning diesel soot into carbon nanotubes.

Just as an aside: there is some truly green nanotechnology: growing nanomaterials in plants – however this will never address industrial production of nanomaterials.

2

Applications of Nanotechnology

Today's nanotechnology harnesses current progress in chemistry, physics, materials science, and biotechnology to create novel materials that have unique properties because their structures are determined on the nanometer scale.

Some of these materials have already found their ways into consumer products, such as sun screens and stain-resistant pants. Others are being intensively researched for solutions to humanity's greatest problems — diseases, clean energy, clean water, etc. Other work is aimed at developing a roadmap for productive nanosystems, in which a path is sought from today's nanotechnology capabilities to advanced future systems in which molecular tools will build useful materials, devices, and complex systems to atomic precision. The products of advanced nanotechnology that will become available in coming decades promise even more revolutionary applications than the products of current and near-term nanotechnology.

With nanotechnology, a large set of materials and improved products rely on a change in the physical properties when the feature sizes are shrunk. Nanoparticles for example take advantage of their dramatically increased surface area to volume ratio.

Their optical properties, *e.g.* fluorescence, become a function of the particle diameter. When brought into a bulk material, nanoparticles can strongly influence the mechanical properties of the material, like stiffness or elasticity.

For example, traditional polymers can be reinforced by nanoparticles resulting in novel materials which can be used as lightweight replacements for metals. Therefore, an increasing societal benefit of such nanoparticles can be expected. Such nanotechnologically enhanced materials will enable a weight reduction accompanied by an increase in stability and an improved functionality. There are many applications of nano technology, few of them are shown here.

MEDICINE

The biological and medical research communities have exploited the unique properties of nanomaterials for various applications (*e.g.*, contrast agents for cell imaging and therapeutics for treating cancer). Terms such as *biomedical nanotechnology*, *bionanotechnology*, and *nanomedicine* are used to describe this hybrid field. Functionalities can be added to nanomaterials by interfacing them with biological molecules or structures.

The size of nanomaterials is similar to that of most biological molecules and structures; therefore, nanomaterials can be useful for both in vivo and in vitro biomedical research and applications. Thus far, the integration of nanomaterials with biology has led to the development of diagnostic devices, contrast agents, analytical tools, physical therapy applications, and drug delivery vehicles.

Diagnostics: Nanotechnology-on-a-chip is one more dimension of lab-on-a-chip technology. Biological tests measuring the presence or activity of selected substances become quicker, more sensitive and more flexible when certain nanoscale particles are put to work as tags or labels.

Magnetic nanoparticles, bound to a suitable antibody, are used to label specific molecules, structures or microorganisms. Gold nanoparticles tagged with short segments of DNA can be used for detection of genetic sequence in a sample. Multicolour optical coding for biological assays has been achieved by embedding different-sized quantum dots into polymeric microbeads. Nanopore technology for analysis of nucleic acids converts strings of nucleotides directly into electronic signatures.

Drug Delivery: The overall drug consumption and side-effects can be lowered significantly by depositing the active agent in the morbid region only and in no higher dose than needed. This highly selective approach reduces costs and human suffering. An example can be found in dendrimers and nanoporous materials. They could hold small drug molecules transporting them to the desired location. Another vision is based on small electromechanical systems; NEMS are being investigated for the active release of drugs. Some potentially important applications include cancer treatment with iron nanoparticles or gold shells.

A targeted or personalized medicine reduces the drug consumption and treatment expenses resulting in an overall societal benefit by reducing the costs to the public health system. Nanotechnology is also opening up new opportunities in implantable delivery systems, which are often preferable to the use of injectable drugs, because the latter frequently display first-order kinetics (the blood concentration goes up rapidly, but drops exponentially over time).

This rapid rise may cause difficulties with toxicity, and drug efficacy can diminish as the drug concentration falls below the targeted range.

Tissue Engineering: Nanotechnology can help to reproduce or to repair damaged tissue. This so called "tissue engineering" makes use of artificially stimulated cell proliferation by using suitable nanomaterial-based scaffolds and growth factors. Tissue engineering might replace today's conventional treatments like organ transplants or artificial implants. On the other hand, tissue engineering is closely related to the ethical debate on human stem cells and its ethical implications.

DEFINITION OF NANOTECHNOLOGY

This definition from the National Nanotechnology Initiative adds the fact that nanotechnology involves certain activities, such as measuring and manipulating nanoscale matter.

Nanotechnology is the science and technology of precisely manipulating the structure of matter at the molecular level. The term nanotechnology embraces many different fields and specialties, including engineering, chemistry, electronics, and medicine, among others, but all are concerned with bringing existing technologies down to a very small scale,

What, actually, do we mean by nanotechnology? The term itself was first coined in 1974 by Tokyo Science University professor Norio Taniguchi, who used it to describe the extension of traditional silicon machining down into regions smaller than one micron. By a now more generally accepted definition, today, nanotechnology is the engineering and fabrication of objects with features smaller than 100 nanometers, or one-tenth of a micron. A micron (lm) is one millionth of a meter–too small for the eye to resolve. A nanometer (nm) is 1 thousandth of a micron–that is to say, really, really tiny. One nanometer is about the size of six carbon atoms aligned, or 10 hydrogen atoms–objects too small to see or image except by the use of very powerful electron or atomic force microscopes.

So we are talking about a molecular scale. The thinnest thing, apparently, that most people are generally aware of is a human hair.

So texts and articles on nanotechnology will tell you that a nanometer is 60 000 times smaller than a human hair is in diameter. Or sometimes the number is 100000; nobody seems to agree. I, personally, have very thin, baby-type hair.

In a laboratory long ago, in a place far away, for the purpose of impressing my daughter, I took one of each of our hairs and placed them under a microscope.

Her hair looked like a cable compared to mine. So I don't use this hackneyed human hair comparison to give people an idea of nanometers. Human hair varies a lot, OK? And mine is almost gone, anyway. The sizes of some fairly well-known biological objects.

A white blood cell is about 10 lm or 10000 nm in diameter. Note that this is actually larger than the interior diameter of the smallest capillaries (8000 nm), so it helps that blood cells are deformable. Bacteria can be as large as a white blood cell, but most are much smaller, on the order of 1 lm in diameter.

Viruses are smaller still, with an upper size range of about 100 nm. Nanofabricated objects have architectural features sizes that are equal to or smaller than the diameter of a virus.

Table. The Sizes of Nanoscale Objects: Nature Versus Fabrication.

Object	Diameter	Object	Diameter
Hydrogen atom	0.1 nm	Buckminsterfullerene (C60)	0.7 nm
Carbon nanotube (single wall)	0.4–1.8 nm	6 carbon atoms aligned	1 nm
DNA	2 nm	Proteins	5–50 nm
CdSe Quantum Dot	2–10 nm	Ribosome	25 nm
Virus	75–100 nm	Semiconductor Chip Features	90 nm or above
Mitochondria	500–1000 nm	Bacteria	1000–10000 nm
Capillary (diameter)	8000 nm	White blood cell	10000 nm

Currently mass-produced semiconductor chips can have circuit elements etched down to 90 nm in diameter. However, this is falling rapidly with new nanolithography techniques, which are already pushing the limit down to around 20 nm, or smaller than the diameter of a ribosome, the organelle within our cells that makes proteins. Carbon nanotubes can have diameters smaller than 2 nm–hence their desirability as potential components in nanoscale chips. Another staple of nanotechnology, the quantum dot, can be manufactured reliably as small as 2 nm in diameter. These enigmatic objects have a variety of uses in biosensors and in electronics.

MEANING OF NANOTECHNOLOGY

Nanotechnology refers to a field of applied science whose theme is the control of matter on an atomic and molecular scale. Generally nanotechnology is approximately 100 nanometers or smaller and involves developing materials or devices within that size. Nanotechnology is a highly multidisciplinary field, drawing from a number of fields such as applied physics, materials science, interface and colloid science, device physics, supramolecular chemistry, self-replicating machines and robotics, chemical engineering, mechanical engineering, biological engineering, and electrical engineering.

When K. Eric Drexler popularized the word 'nanotechnology' in the 1980's, he was talking about building machines on the scale of molecules, a few nanometres wide—motors, robot arms, and even whole computers, far smaller than a cell. Drexler spent the next ten years describing and analysing these incredible devices, and responding to accusations of science fiction. Meanwhile, mundane technology was developing the ability to build simple structures on a molecular scale. As nanotechnology became an accepted concept, the meaning of the word shifted to encompass the simpler kinds of nanometre-scale technology. The National Nanotechnology Initiative was created to fund this kind of nanotech: their definition includes anything smaller than 100 nanometres with novel properties.

Much of the work being done today that carries the name 'nanotechnology' is not nanotechnology in the original meaning of the word. Nanotechnology, in its traditional sense, means building things from the bottom up, with atomic precision.

This theoretical capability was envisioned as early as 1959 by the renowned physicist Richard Feynman. I want to build a billion tiny factories, models of each other, which are manufacturing simultaneously. The principles of physics, as far as I can see, do not speak against the possibility of maneuvering things atom by atom. It is not an attempt to violate any laws; it is something, in principle, that can be done; but in practice, it has not been done because we are too big.Richard Feynman, Nobel Prize winner in physics

Based on Feynman's vision of miniature factories using nanomachines to build complex products, advanced nanotechnology (sometimes referred to as molecular manufacturing) will make use of positionally-controlled mechanochemistry guided by molecular machine systems. Formulating a roadmap for development of this kind of nanotechnology is now an objective of a broadly based technology roadmap project led by Battelle.

Shortly after this envisioned molecular machinery is created, it will result in a manufacturing revolution, probably causing severe disruption. It also has serious economic, social, environmental, and military implications.

GENERATIONS OF NANOTECHNOLOGY

Mihail Roco of the National Nanotechnology Initiative has described *four generations* of nanotechnology development. The current era, as Roco depicts it, is that of passive nanostructures, materials designed to perform one task.

The second phase, which we are just entering, introduces active nanostructures for multitasking; for example, actuators, drug delivery devices, and sensors. The third generation is expected to begin emerging around 2010 and will feature nanosystems with thousands of interacting components.

A few years after that, the first integrated nanosystems, functioning much like a mammalian cell with hierarchical systems within systems, are expected to be developed.

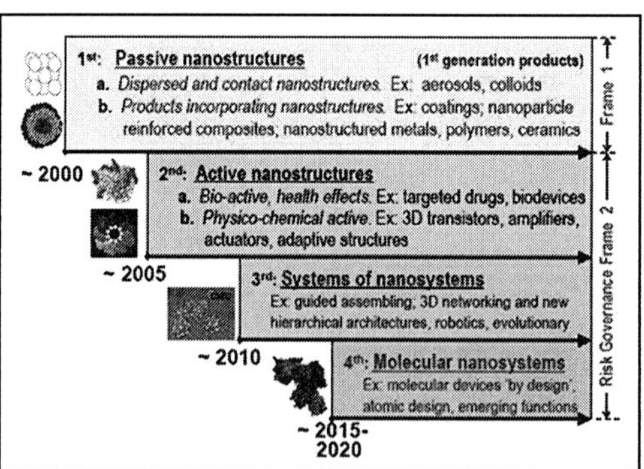

Some experts may still insist that nanotechnology can refer to measurement or visualization at the scale of 1-100 nanometres, but a consensus seems to be forming around the idea that *control and restructuring of matter* at the nanoscale is a necessary element.

CRN's definition is a bit more precise than that, but as work progresses through the four generations of nanotechnology leading up to molecular nanosystems, which will include molecular manufacturing, we think it will become increasingly obvious that "engineering of functional systems at the molecular scale" is what nanotech is really all about.

CONFLICTING DEFINITIONS

Unfortunately, conflicting definitions of nanotechnology and blurry distinctions between significantly different fields have complicated the effort to understand the differences and develop sensible, effective policy. The risks of today's nanoscale technologies (nanoparticle toxicity, etc.) cannot be treated the same as the risks of longer-term molecular manufacturing (economic disruption, unstable arms race, etc.).

It is a mistake to put them together in one basket for policy consideration each is important to address, but they offer different problems and will require different solutions. As used today, the term nanotechnology usually refers to a broad collection of mostly disconnected fields. Essentially, anything sufficiently small and interesting can be called nanotechnology. Much of it is harmless. For the rest, much of the harm is of familiar and limited quality. But as we will see, molecular manufacturing will bring unfamiliar risks and new classes of problems.

GENERAL-PURPOSE TECHNOLOGY

Nanotechnology is sometimes referred to as a *general-purpose technology*. That's because in its advanced form it will have significant impact on almost all industries and all areas of society. It will offer better built, longer lasting, cleaner, safer, and smarter products for the home, for communications, for medicine, for transportation, for agriculture, and for industry in general.

Imagine a medical device that travels through the human body to seek out and destroy small clusters of cancerous cells before they can spread. Or a box no larger than a sugar cube that contains the entire contents of the Library of Congress. Or materials much lighter than steel that possess ten times as much strength. — U.S. National Science Foundation.

DUAL-USE TECHNOLOGY

Like electricity or computers before it, nanotech will offer greatly improved efficiency in almost every facet of life. But as a general-purpose technology, it will be *dual-use*, meaning it will have many commercial uses and it also will have many military uses—making far more powerful weapons and tools of surveillance. Thus it represents not only wonderful benefits for humanity, but also grave risks.

A key understanding of nanotechnology is that it offers not just better products, but a vastly improved manufacturing *process*. A computer can make copies of data files—essentially as many copies as you want at little or no cost.

It may be only a matter of time until the building of products becomes as cheap as the copying of files. That's the real meaning of nanotechnology, and why it is sometimes seen as "the next industrial revolution."

My own judgement is that the nanotechnology revolution has the potential to change America on a scale equal to, if not greater than, the computer revolution. The power of nanotechnology can be encapsulated in an apparently simple device called a *personal nanofactory* that may sit on your countertop or desktop. Packed with miniature chemical processors, computing, and robotics, it will produce a wide-range of items quickly, cleanly, and inexpensively, building products directly from blueprints.

EXPONENTIAL PROLIFERATION IN NANOTECHNOLOGY

Nanotechnology not only will allow making many high-quality products at very low cost, but it will allow making new nanofactories at the same low cost and at the same rapid speed. This unique (outside of biology, that is) ability to reproduce its own means of production is why nanotech is said to be an *exponential* technology. It represents a manufacturing system that will be able to make more manufacturing systems—factories that can build factories—rapidly, cheaply, and cleanly. The means of production will be able to reproduce exponentially, so in just a few weeks a few nanofactories conceivably could become *billions*. It is a revolutionary, transformative, powerful, and potentially very dangerous—or beneficial—technology.

How soon will all this come about? Conservative estimates usually say 20 to 30 years from now, or even much later than that. However, CRN is concerned that it may occur sooner, quite possibly within the next decade. This is because of the rapid progress being made in enabling technologies, such as optics, nanolithography, mechano-chemistry and 3D prototyping.

If it does arrive that soon, we may not be adequately prepared, and the consequences could be severe.

We believe it's not too early to begin asking some tough questions and facing the issues:

- Who will own the technology?
- Will it be heavily restricted, or widely available?
- What will it do to the gap between rich and poor?
- How can dangerous weapons be controlled, and perilous arms races be prevented?

Many of these questions were first raised over a decade ago, and have not yet been answered. If the questions are not answered with deliberation, answers will evolve independently and will take us by surprise; the surprise is likely to be unpleasant. It is difficult to say for sure how soon this technology will mature, partly because it's possible (especially in countries that do not have open societies) that clandestine military or industrial development programmes have been going on for years without our knowledge. We cannot say with certainty that full-scale nanote-chnology will not be developed with the next ten years, or even five years. It may take longer than that, but prudence—and possibly our survival—demands that we prepare now for the earliest plausible development scenario.

CONTEXT OF NANOTECHNOLOGY

Nanotechnology is the identification, application and use of novel behaviour that occurs at the nanoscale to solve real-world problems. The discipline requires a breadth of understanding that is much wider than just the equations and scientific principles that underlie that behaviour. This introductory course gives an overview of the current state of nanotechnology as well as introducing the implications of these new technologies for safety, regulation, and innovation. The course provides an overview of the societal and environmental implications of nanotechnology.

SCIENTIFIC REVOLUTIONS

The development of man is marked by technological breakthroughs. So important are they that the technologies give their names to the successive epochs: the Stone Age, the Bronze Age, the Iron Age, rather than the age of hunting, pastoralism, agriculture, urbanization etc.

The most significant change in our way of life during the last two or three millennia was probably that brought about by the Industrial Revolution that began in Britain around the middle of the 18th century; by the middle of the 19th century it was in full swing in Britain and, at first patchily, but later rapidly, lsewhere in Europe and North America. This in turn was replaced by the Information Revolution, marked by unprecedented capabilities in the gathering, storage, retrieval and analysis of information, and heavily dependent upon the high-speed electronic digital computer.

We are still within that epoch, but the next revolution already appears to be on the horizon, and it is thought that it will be the Nano Revolution. There are a couple of things worth noting about these revolutions. Firstly, the exponential growth in capabilities. This is sometimes quite difficult to accept, because an exponential function is linear if examined over a sufficiently small interval, and if the technology unfolds over several generations, individual perceptions tend to be strongly biased towards linearity.

Nevertheless, empirical examination of available data shows that exponential development is the rule, although it does not continue indefinitely, but eventually levels off. Secondly, very often a preceding technological breakthrough provided the key to a successive one. For example, increasing skill and knowledge in working iron was crucial to the success of the steam power and steel that were the hallmarks of the Industrial Revolution, which ultimately developed the capability for mass production of the very large-scale integrated electronic circuits needed for realizing the Information Revolution.

Why do people think that the next technological revolution will be that of nanotechnology? Because once we have mastered the technology, the advantages of making things "at the bottom" will be so overwhelming it will rapidly dominate all existing ways of doing things. Once ironmaking and working had been mastered, no one would have considered making large, strong objects out of bronze; no one uses a slide rule now that electronic calculators are available; and so forth.

THE ADVANTAGES OF NANOTECHNOLOGY

They follow from miniaturization, novel combinations of properties, and a universal fabrication technology. Typical of the benefits of miniaturization is the cell phone. The concept was developed in the 1950s, but the necessary circuitry would have occupied a large multistorey building using the technology of the day. Materials made with carbon nanotubes can be light and very strong, and transparent and electrically conducting. Universal fabrication, based on assemblers would enable most artefacts required by humans to be made out of acetylene and a source of energy.

REALIZING OF THE NANO REVOLUTION

Miniaturization of circuitry is lready far advanced. Components and chips can now be made with features in the size range of tens of nanometres. The World Wide Web would be scarcely conceivable without the widespread dissemination of powerful personal computers enabled by mass-produced integrated circuits. Materials based on carbon nanotubes are still very much at the experimental stage. Nevertheless, prototypes have been made and the difficulties look to be surmountable. Assembly-based nanofacture seems still to be some way in the future. To demonstrate feasibility, computer simulations are generally adduced, together with biological systems. Nevertheless, actual experiments demons-trating assembly with atomic precision are still in a primitive state.

What might the benefits be: Reports published during the last few years are typically euphoric about nanotechnology and all the benefits it will bring. Many of the examples are, however, of a relatively trivial nature and do not seem to represent sufficient breakthrough novelty to constitute a revolution. Thus, we already have nanostructured textiles that resist staining, self-cleaning glass incorporating nanoparticulate photocatalysts capable of decomposing dirt; nanoparticle-based sun creams that effectively filter out ultraviolet light without scattering it and are therefore transparent; even lighter and stronger tennis racquets made with carbon fibre or even carbon nanotube composites; and so forth. None of these developments can be said to be truly revolutionary in terms of impact on civilization.

The Industrial Revolution was very visible because of the colossal size of its products: gigantic bridges, gigantic steamships and, most gigantic of all if the entire network is considered as a single machine, the railway. And the steel for these constructions was produced in gigantic works; a modern chemical plant or motor-car factory may cover the area of a medium-sized town.

In sharp contrast, the products of nanotechnology are, by definition, very small. Individual assemblers would be invisible to the naked eye. But of course the products of the assemblers would be highly visible and pervasive—such as ultralight strong materials from which our built environment would be constructed.

Microprocessors grading into nanoprocessors are a manifestation of indirect nanotechnology, responsible for the ubiquity of internet servers and cellular telephones. The impact of these information processors is above all due to their very high-speed operation, rather than any particular sophistication of the algorithms governing them.

Most tasks, ranging from the diagnosis of disease to surveillance, involve pattern recognition, something that our brains can accomplish swiftly and seemingly effortlessly for a while, but which require huge numbers of logical steps when reduced to a form suitable for a digital processor. Sanguine observers predict that despite the clumsiness of this "automated reasoning", ultimately artificial thinking will surpass that of humans—this is Kurzweil's "singularity". Others predict that it will never happen. To be sure, the singularity is truly revolutionary, but is as much a product of the Information Revolution as of the Nano Revolution, even though the latter provides the essential enabling technology.

Conceptual nanotechnology implies scrutinizing the world from the viewpoint of the atom or molecule. In medicine this amounts to finding the molecular basis of disease, which has been underway ever since biochemistry became established, and which now encompasses all aspects of disease connected with the DNA molecule and its relatives. There can be little doubt about the tremendous advance of knowledge that it represents. It, however, is part of the more general scientific revolution that began in the European universities founded from the 11th century onwards—and which is so gradual and ongoing that it never really constitutes a perceptible revolution. Furthermore, it is always necessary to counterbalance the reductionism implicit in the essentially analytical atomic viewpoint by insisting on a synthetic systems approach at the same time.

Nanotechnology carried through to Productive Nanosystems could achieve this, because the tiny artefacts produced by an individual assembler have somehow to be transformed into something macroscopic enough to be serviceable for mankind.

Can nanotechnology help to solve the great and pressing problems of contemporary humanity? Although, if ranked, there might be some debate about the order, most people would include rapid climate change, environmental degradation, depletion of energy, unfavourable demographic trends, insufficiency of food, and nuclear proliferation among the biggest challenges.Seen from this perspective, nanotechnology is the continuation of technological progress, which might ultimately be revolutionary if the quantitative change becomes big enough to rank as qualitative.

For example, atom-by-atom assembly of artefacts implies discarded ones can be disassembled according to a similar principle, hence the problem of waste vanishes. More advanced understanding at the nanoscale should finally allow us to create artificial energy harvesting systems, hence the potential penury of energy disappears.

If the manufacture of almost everything becomes localized, the transport of goods should dwindle to practically nothing. Localized energy production would have a similar effect. However, the achievement of this state of affairs depends on the advent of the personal nanofactory, or something resembling it, which is by no means inevitable. Perhaps the nanobot is somewhat closer to realization. ould indefatigably circulating nanobots inside our bodies enable our lives to be extended almost indefinitely? And what would be the consequences?

NANOSCIENCE

Is there a need for this term? Sometimes it is defined as "the science underlying nanotechnology". But this really is biology, chemistry and physics—or "molecular sciences". It is the technology of designing and making functional objects at the nanoscale that is new; science has long been working at this scale, and below. No one is arguing that fundamentally new physics emerges at the nanoscale; rather, it is the new combinations of phenomena manifesting themselves at that scale that constitute the new technology. The term "nanoscience" therefore appears to be wholly superfluous if it is used in this sense. As a synonym of conceptual nanotechnology, however, it does have a useful meaning: the science of mesoscale approximation. The description of a protein as a string of amino acids is a good example. At the mesoscale, one does not need to enquire into details of the internal structure of the amino acids.

MOTIVATION FOR NANOTECHNOLOGY

We consider approximately of the causes why one might prefer to make matters identical belittled, considering nanotechnology along the "materials, devices, and arrangements".

MATERIALS

Most of the materials around us are composites. Natural materials such as wood are highly structured and built upon very sophisticated principles. The basic structural unit is cellulose, which is a polymer of the sugar glucose, but cellulose on its own makes a floppy fabric, hence to give it strength and rigidity it must be glued together into a rigid matrix.

This is accomplished by the complex multiring aromatic molecule lignin. The design principle is therefore akin to that of reinforced concrete: Steel rods strengthen what is itself a composite of gravel and cement.

The principle of combining two or more pure substances with distinctly different properties in order to create a composite material that combines the desirable properties of each to create a multifunctional substance has been refined over millennia, presumably mostly by trial and error. Typically, the results are, to a first approximation, additive. Thus we might write a sum of materials and their properties like

Cellulose	High Tensile Strength	Self-repellent
+ Lignin	Weak	Sticky
4= Wood	Strong	Cohesive

Empirical knowledge is used to choose useful combinations, in which the desirable properties dominate—one might have ended up with a weak and repellent material.

The vast and growing accumulation of empirical knowledge, now backed up and extended by fundamental knowledge of the molecular-scale forces involved, usually allow appropriate combinations to be chosen. The motif of strong fibres embedded in a sticky matrix is very widely exploited, other examples being glass fibre- and carbon fibre-reinforced polymers. Essentially, the contribution of nanotechnology to this effort is simply to take it to the ultimate level, in the spirit of “shaping the world atom-by-atom”.

Rather like the chemist trying to synthesize an elaborate multifunctional molecule, the materials nanotechnologist aims to juxtapose different atoms to achieve multifunctionality.

This approach is known as mechanosynthetic chemistry or, in its large-scale industrial realization, as molecular manufacturing. The famous experiment of Schweizer and Eigler, in which they rearranged xenon atoms on a nickel surface to form the logo “IBM”, represented a first step in this direction.

Since then, there has been intensive activity in the area, but it still remains uncertain to what extent arbitrary combinations of atoms can be assembled disregarding chemical concepts, and whether the process can ever be scaled up to provide macroscopic quantities of materials.

Most of the recognized successes in nanomaterials so far have been not in the creation of totally new materials through mechanosynthesis, but in the more prosaic world of blending.

For example, one adds hard particles to a soft polymer matrix to create a hard, abrasion-resistant coating. As with atomically-based mechanosynthesis, the results are, to a first approximation, additive. Thus we might again write a sum like

	Polypropylene	Flexible	Transparent
+	Titanium dioxide	Rigid	Opaque
=	Thin film coating (paint)	Flexible	Opaque

This is not actually very new. Paint, a blend of pigment particles in a matrix has been manufactured for millennia. What is new is the detailed attention paid to the nanoparticulate additive. Its properties can now be carefully tailored for the desired application. If one of components is a recognized nanosubstance—a nanoparticle or nanofibre, for example—it seems to be acceptable to describe the blend as a nanomaterial.

The biggest range of applications for such nanocomposites is in thin film coatings—in other words paint. Traditional pigments may comprise granules in the micrometre size range; grinding them a little bit more finely turns them into nano-objects. Compared with transparent varnish, paint then combines the attribute of protection from the environment with the attribute of colour. The principle can obviously be extended practically ad libitum: by adding very hard particles to confer abrasion resistance; metallic particles to confer electrical conductivity; tabular particles to confer low gas permeability, and so on.

Two relatively old products even today constitute the bulk of the so-called nanotechnology industry: carbon black carbon particles ranging in size from a few to several hundred nanometres) added to the rubber tyres for road vehicles as reinforcing filler; and crystals of silver chloride, silver bromide and silver iodide ranging in size from tens of nanometres to micrometres, which form the basis of conventional silver halide-based photography.

Why Nanoadditives: Since it is usually more expensive to create nanosized rather than microsized matter, one needs to justify the expense of downscaling. As matter is divided ever more finely, certain properties become qualitatively different.

For example, the optical absorption spectrum of silicon vapour is quite different from that of a silicon crystal, even though the vapour and crystal are chemically identical. When a crystal becomes very small, the melting point falls, there may be a lattice contraction —these are well understood consequences of Laplace's law, and may be very seful for facilitating a sintering process. If the radius of the crystal is smaller than the Bohr adius of the electron in the bulk solid, the electron is confined and has a higher energy than its bulk counterpart. The optical absorption and fluorescent emission spectra shift to higher energies.

Hence, by varying the crystal radius, the optical absorption and emission wavelengths can be tuned. Chemists have long known that heterogeneous catalysts are more active if they are more finely divided. This is a simple consequence of the fact that the reaction takes place at the interface between the solid catalyst and the rest of the reaction medium. For a given mass, the finer the division, the greater the surface area. This is not in itself a qualitative change, although in an industrial application there may be a qualitative transition from an uneconomic to an economic process. Our planet has an oxidizing atmosphere, and has had one probably for at least 2000 million years.

This implies that most metals, other than gold, platinum and so forth, will be oxidized. Hence, many kinds of metallic nanoparticles will not be stable in nature. Carbon-based materials, especially fullerenes in carbon nanotubes, are often considered to e the epitome of a nanomaterial. Carbon has long been an intriguing element because of the enormous differences between its allotropes of graphite and diamond. The carbon nanomaterials are based on another new form, graphene.

Devices: A device turns something into something else. Synonyms are machine, automaton, transducer, encoder, and so forth. Possible motivations for miniaturizing a device are:

- *Economizing on Material*: If one can accomplish the same function with less material, the device should be cheaper, which is often a desirable goal—provided that it is not more expensive to make. In the macroscopic world of mechanical engineering, if the material osts are disregarded, it is typically more expensive to make something very small; for example, a watch is more expensive than a clock, for equivalent timekeeping precision. On the other hand, when things become very large, as in the case of the clock familiarly known as Big Ben for example, costs again start to rise, because special machinery may be needed to assemble the components, and so on.
- *Performance may be Enhanced by Reducing the Size*: This is actually quite rare. For most *micro electromechanical systems* (MEMS) devices, such as accelerometers, performance is degraded by downscaling, and the actual size of the devices currently mass-produced for actuating automotive airbags represents a compromise between economy of material, not taking up too much space nor weighing two much, the and still-acceptable performance.
- *Functionality may be Enhanced by Reducing the Size:* Using the same example as in the previous item, it would not be practicable to equip mass-produced automobiles with macroscopic accelerometers with a volume of about 1 litre and weighing several kilograms. Another example is cellular telephony. A similar consideration applies to implantable biosensors for monitoring clinical parameters in a patient. In other words, miniaturization increases accessibility.

Systems: The essence of a system is that it cannot be usefully decomposed into its constituent parts. Two or more objects constitute a system if the following conditions are satisfied:

- One can talk meaningfully of the behaviour of the whole of which they are the only parts

- The behaviour of each part can affect the behaviour of the whole
- The way each part behaves and the way its behaviour affects the whole depends on the behaviour of at least one other part
- No matter how one subgroups the parts, the behaviour of each subgroup will affect the whole and depends on the behaviour of at least one other subgroup.

Typically, a single nanodevice is complex enough to be considered a system, hence a "nanosystem" generally signifies a system whose components are nanoscale devices.

An example of a system that can truly be called "nano" is the foot of the gecko, many species of which can run up vertical walls and across ceilings.

Their feet are hierarchically divided into tens of thousands of minute pads that allow a large area of conformal contact with irregular surfaces. The adhesive force is provided by the Lifshitz-van der Waals interaction, normally considered to be weak and short range, but additive and hence sufficiently strong in this embodiment if here are enough points of contact.

Attempts to mimic the foot with a synthetic nanostructure ave only had very limited success, because the real foot is living and constantly adjusted to maintain the close range conformal contact needed for the interaction to be sufficiently strong to bear the weight of the creature.

Issues in miniaturization: Considering the motor-car as a transducer of human desire into translational motion, it is obvious that the nanoautomobile would be useless for transporting anything other than nanoobjects.

The main contribution of nanotechnology to the automotive industry is in providing miniature sensors for process monitoring in various parts of the engine and air quality monitoring n the saloon; additives in paint giving good abrasion resistance, possibly self-cleaning functionality, and perhaps novel aesthetic effects; new ultrastrong and ultralightweight composites incorporating carbon nanotubes for structural parts; sensors embedded in the chassis nd bodywork to monitor structural health; and so forth.

Scaling Up: In other cases, scaling performance up to the level of human utility is simply a matter of massive parallelization.

Nanoreactors synthesizing a medicinal drug simply need to work in parallel for a reasonably short time to generate enough of the compound for a therapeutically useful dose.

With information processors, the problem is the user interface: a visual display screen must be large enough to display a useful amount of information, a keyboard for entering instructions and data must be large enough for human fingers, and so forth.

OTHER MOTIVATIONS

The burgeoning worldwide activity in nanotechnology cannot be explained purely as a rational attempt to exploit "room at the bottom". Two other important human motivations are doubtless also playing a role. One it is simply "it hasn't been done before"—the motivation of the mountaineer ascending a peak previously untrodden.

The other is the perennial desire to conquer nature. Opportunities for doing so at the familiar macroscopic scale have become very limited, partly because so much has already been donein Europe, for example, there are hardly any marshes left to drain or rivers left to dam, two of the most typical arenas for conquering natureand partly because the deleterious effects of such conquest are now far more widely recognized, and the few remaining undrained marshes and undammed rivers are likely nowadays to be legally protected nature reserves. But the world at the bottom is uncontrolled and largely unexplored. Finally, the space industry has a constant and heavily pressing requirement for making payloads as small and lightweight as possible. Nanotechnology is ideally suited to this end userprovided nanomaterials, devices and systems can be made sufficiently reliable.

FILTRATION TECHNIQUES OF CHEMISTRY AND ENVIRONMENT

Chemical catalysis and filtration techniques are two prominent examples where nanotechnology already plays a role. The synthesis provides novel materials with tailored features and chemical properties: for example, nanoparticles with a distinct chemical surrounding (ligands), or specific optical properties.

In this sense, chemistry is indeed a basic nanoscience. In a short-term perspective, chemistry will provide novel "nanomaterials" and in the long run, superior processes such as "self-assembly" will enable energy and time preserving strategies.

In a sense, all chemical synthesis can be understood in terms of nanotechnology, because of its ability to manufacture certain molecules. Thus, chemistry forms a base for nanotechnology providing tailor-made molecules, polymers, etcetera, as well as clusters and nanoparticles.

Benefits of Chemical catalysis: Chemical catalysis benefits especially from nanoparticles, due to the extremely large surface to volume ratio. The application potential of nanoparticles in catalysis ranges from fuel cell to catalytic converters and photocatalytic devices. Catalysis is also important for the production of chemicals. Platinum nanoparticles are now being considered in the next generation of automotive catalytic converters because the very high surface area of nanoparticles could reduce the amount of platinum required.

However, some concerns have been raised due to experiments demonstrating that they will spontaneously combust if methane is mixed with the ambient air. Ongoing research at the Centre National de la Recherche Scientifique (CNRS) in France may resolve their true usefulness for catalytic applications. Nanofiltration may come to be an important application, although future research must be careful to investigate possible toxicity.

Filtration: A strong influence of nanochemistry on waste-water treatment, air purification and energy storage devices is to be expected. Mechanical or chemical methods can be used for effective filtration techniques. One class of filtration techniques is based on the use of membranes with suitable hole sizes, whereby the liquid is pressed through the membrane.

Nanoporous membranes are suitable for a mechanical filtration with extremely small pores smaller than 10 nm ("nanofiltration") and may be composed of nanotubes. Nanofiltration is mainly used for the removal of ions or the separation of different fluids. On a larger scale, the membrane filtration technique is named ultrafiltration, which works down to between 10 and 100 nm. One important field of application for ultrafiltration is medical purposes as can be found in renal dialysis. Magnetic nanoparticles offer an effective and reliable method to remove heavy metal contaminants from waste water by making use of magnetic separation techniques.

Using nanoscale particles increases the efficiency to absorb the contaminants and is comparatively inexpensive compared to traditional precipitation and filtration methods. Some water-treatment devices incorporating nanotechnology are already on the market, with more in development. Low-cost nanostructured separation membranes methods have been shown to be effective in producing potable water in a recent study.

ADVANCED ENERGY IN NANOTECHNOLOGY

The most advanced nanotechnology projects related to energy are: storage, conversion, manufacturing improvements by reducing materials and process rates, energy saving (by better thermal insulation for example), and enhanced renewable energy sources.

Reduction of Energy Consumption: A reduction of energy consumption can be reached by better insulation systems, by the use of more efficient lighting or combustion systems, and by use of lighter and stronger materials in the transportation sector.

Currently used light bulbs only convert approximately 5 per cent of the electrical energy into light. Nanotechnological approaches like light-emitting diodes (LEDs) or quantum caged atoms (QCAs) could lead to a strong reduction of energy consumption for illumination.

Increasing the Efficiency of Energy Production: Today's best solar cells have layers of several different semiconductors stacked together to absorb light at different energies but they still only manage to use 40 per cent of the Sun's energy. Commercially available solar cells have much lower efficiencies (15-20 per cent).

Nanotechnology could help increase the efficiency of light conversion by using nanostructures with a continuum of bandgaps. The degree of efficiency of the internal combustion engine is about 30-40 per cent at the moment. Nanotechnology could improve combustion by designing specific catalysts with maximized surface area. In 2005, scientists at the University of Toronto developed a spray-on nanoparticle substance that, when applied to a surface, instantly transforms it into a solar collector.

The Use of More Environmentally Friendly Energy Systems: An example for an environmentally friendly form of energy is the use of fuel cells powered by hydrogen, which is ideally produced by renewable energies. Probably the most prominent nanostructured material in fuel cells is the catalyst consisting of carbon supported noble metal particles with diameters of 1-5 nm. Suitable materials for hydrogen storage contain a large number of small nanosized pores.

Therefore many nanostructured materials like nanotubes, zeolites or alanates are under investigation. Nanotechnology can contribute to the further reduction of combustion engine pollutants by nanoporous filters, which can clean the exhaust mechanically, by catalytic converters based on nanoscale noble metal particles or by catalytic coatings on cylinder walls and catalytic nanoparticles as additive for fuels.

Recycling of Batteries: Because of the relatively low energy density of batteries the operating time is limited and a replacement or recharging is needed. The huge number of spent batteries and accumulators represent a disposal problem. The use of batteries with higher energy content or the use of rechargeable batteries or supercapacitors with higher rate of recharging using nanomaterials could be helpful for the battery disposal problem.

RESEARCH IN ENCOURAGING ENVIRONMENTALLY BENEFICIAL NANOTECHNOLOGY

The role of research in encouraging environmentally beneficial nanotechnology. However, there is also a need for research focusing on adverse effects of nanotechnology, monitoring instruments, and control methods to prevent adverse effects. Much of the research on effects is either done or paid for by the government. nanotechnology presents a variety of unique challenges. It is not clear whether there are good ways to detect nanotechnology materials in the ambient environment, or, if they cannot be detected, whether there are ways environmental controls can be imposed.

It is not clear which, if any, of the usual control methods will be effective in dealing with nanotechnology materials. Following these avenues often leads back to the primary importance of prevention. The only way to deal with potential adverse effects of nanotechnology in most cases will be to design the product or tailor the use of the material so that the nanotechnology material does not get into the environment or the human body in the first place.

Rejeski has suggested that EPA needs an advanced research capability similar to the military's Defence Advanced Research Projects Agency to develop breakthrough solutions for the "really hard problems". EPA's track record with research of this kind is not good, but creation of a new office with incentives to attract outstanding researchers could change this.

It will remain a continuing challenge to identify and analyse the adverse effects of nanotechnology. With government funds,some new institutions at universities already have been established to do research on these effects. Much of the research on effects is cross-disciplinary—a major handicap in enlisting the discipline-oriented universities to undertake the research.

It also is a handicap in enlisting individual researchers who may not get academic rewards for working on interdisciplinary, applied problems. The question of how to create incentives for the private sector to conduct testing and do research on nanotechnology adverse effects is a different kind of institutional challenge. Holding manufacturers responsible for the adverse effects of their products is clearly a major incentive for such testing and research.

There is broad agreement that current government spending to determine the health and safety effects of nanotechnology is inadequate.At a Nov. 17, 2005, hearing before the House Committee on Science, industry, environmental and research organizations called for federal spending of at least $100 million annually—more than three times current expenditures—on health and safety research. Some witnesses also called for a more actively managed, strategically targeted, and carefully coordinated approach to determine what adverse effects, if any, nanotechnology may create.

Recently, the Project on Emerging Nanotechnologies at the Woodrow Wilson Center released a comprehensive inventory of nanotechnology-related environmental, health and safety research. The inventory covered information on 208 research projects, in 6 countries and regions, accounting for $38 million of research annually. 169 projects, accounting for $27 million,were in the United States. However, within the United States, only $6 million of federal research was considered to be of "high relevance" to determining the environmental, health and safety implications of engineered nanomaterials.

AC ELECTROKINETICS: APPLICATIONS FOR NANOTECHNOLOGY

The phenomena of dielectrophoresis and electrorotation, collectively referred to as AC electrokinetics, have been used for many years to study, manipulate and separate particles on the cellular (1 mm or more) scale. However, the technique has much to offer the expanding field of nanotechnology, that is the precise manipulation of particles on the nanometre scale. The principles of AC electrokinetics for particle manipulation, review the current state of AC Electrokinetic techniques for the manipulation of particles on the nanometre scale, and consider how these principles may be applied to nanotechnology.

There are two principal approaches to manufacturing on the molecular scale: self-assembly of machines from basic chemical building blocks (the "bottom up" approach) and assembly by manipulating components with much larger devices (the "top down" approach).

Whilst the former is considered to be an ideal through which nanotechnology will ultimately be implemented, the latter approach more readily achievable using current technology. Examples of this include the manipulation of single xenon atoms on a silicon surface and the trapping of single 3nm colloidal particles from solution using electrostatic methods. Another group of methods with potential for use in top-down manufacturing — collectively referred to as AC electrokinetics — will be discussed.

AC Electrokinetic techniques such as dielectrophoresis and electrorotation have been utilised for many years for the manipulation, separation and analysis of cellular-scale particles. The phenomenon occurs due to the interaction of induced dipoles with electric fields, and can be used to exhibit a variety of motions including attraction, repulsion and rotation by changing the nature of the dynamic field. In many ways, these forces may be viewed as an electrostatic equivalent to optical tweezers and optical spanners in that they exert translational and rotational forces on a body due to the interaction between a body and an imposed field gradient.

Recent advances in semiconductor manufacturing technology have enabled researchers to develop electrodes for manipulating macromolecules as small as 9kDa using both attractive and repulsive AC electrostatic forces, and to concentrate 14nm beads from solution. Trapping of single particles such as viruses and 93nm-diameter latex spheres in contactless potential energy wells has also been demonstrated.

AC Electrokinetics offers advantages over scanning-probe methods of nanoparticle manipulation in that the equipment used is simple, cheap and has no moving parts, relying entirely on the electrostatic interactions between the particle and dynamic electric field.

Furthermore, there is theoretical evidence that as manufacturing technology further improves, single particles considerably smaller than presently studied using AC Electrokinetics may be manipulated.

Ultimately, such a technology will have obvious applications for the manipulation of single molecules.

Recent studies have shown that the trapping efficiency of planar electrode arrays is dependent on a number of factors including the magnitude and dimensions of the electric field and the radius of the particle to be trapped. In this paper we will consider the ways in which AC Electrokinetics can benefit nanotechnology, and the constraints on the technique due to factors such as Brownian motion, heating of the medium, and electrode dimensions as the electrode array is miniaturised to the nanometre scale.

DIELECTROPHORESIS

Consider a dielectric particle suspended in a spatially non-uniform electric field such as that shown in figure. The applied field induces a dipole in the particle; the interaction of the induced dipole with the electric field generates a force. Due to the presence of a field gradient, these forces are not equal and there is a net movement. If the particle is more conductive than the medium around it, the dipole aligns with the field and the force acts up the field gradient towards the region of highest electric field. If the particle is less polarisable than the medium, the dipole aligns against the field and the particle is repelled from regions of high electric field. The force is dependent on the induced dipole, and is unaffected by the direction of the electric field, responding only to the field gradient. Since the alignment of the field is irrelevant, this force can also be generated in AC fields which has the advantage of reducing any *electrophoretic* force to zero.

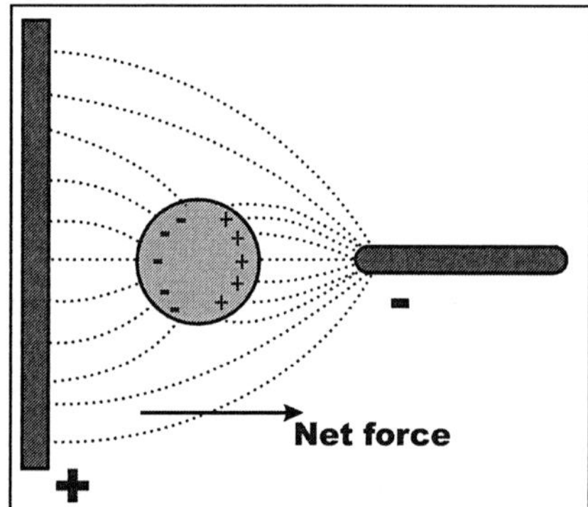

Fig: A schematic of a Polarisable Particle Suspended within a Point-Plane Electrode System.

When the particle polarises, the interaction between the dipolar charges with the local electric field produces a force. Due to the inhomogeneous nature of the electric field, the force is greater in the side facing the point than that on the side facing the plane, and there is net motion towards the point electrode. This effect is called positive dielectrophoresis. If the particle is less polarisable than the surrounding medium, the dipole will align counter to the field and the particle will be repelled from the high field regions, called negative dielectrophoresis.

This effect was first termed Dielectrophoresis by Pohl. The dielectrophoretic force,, acting on a homogeneous, isotropic dielectric sphere, is given by:

where r is the particle radius, is the permittivity of the suspending medium, ∇ is the Del vector operator, E is the *rms.* electric field and, the real part of the Clausius-Mossotti factor, given by: where and are the complex permittivities of the medium and particle respectively, and with σ the conductivity, ε the permittivity and ω the angular frequency of the applied electric field. The frequency-dependence of, indicates that the force on the particle varies with the frequency. The magnitude of, also varies depending on whether the particle is more or less polarisable than the medium. If, is positive, then particles move to regions of highest field strength (positive dielectrophoresis); the converse is negative dielectrophoresis where particles are repelled from these regions. By careful construction of the electrode geometry which creates the electric field, it is possible to create electric field morphologies so that potential energy minima are bounded by regions of increasing electric field strengths.

In such electrodes, particles experiencing positive dielectrophoresis are attracted to the regions of highest electric field (typically the electrode edges, particularly where adjacent electrodes are close), whilst particles experiencing negative dielectrophoresis are trapped in isolated field minima.

ELECTROROTATION

If a polarisable particle is suspended in a rotating electric field, the induced dipole will form across the particle and rotate in synchrony with the field. However, if the angular velocity of the field is sufficiently large, the time taken for the dipole to form (the relaxation time of the dipole) becomes significant and the dipole will lag behind the field.

If the electric field E rotates sufficiently quickly, the induced dipole M will lag behind the electric field by an angle related to the time taken for the dipole to form (the relaxation time). The interaction between the electric field and the lagging dipole induces a torque G in the particle, causing the particle to rotate. This effect is known as *electrorotation*.

The general equation for time-averaged torque G experienced by a spherical polarisable particle of radius r suspended in a rotating electric field E is given by:

where $Im[K(\omega)]$ represents the imaginary component of the Clausius-Mossotti factor. The minus sign indicates that the dipole moment lags the electric field. When viscous drag is accounted for, the rotation rate R(ω) of the particle is given.

Note that there are two significant differences between the expression for torque as shown in equation.

Firstly, the relationship with the electric field is as a function of the square of the electric field rather than of the *gradient* of the square of the electric field. Secondly, the torque depends on the imaginary rather than the real part of the Clausius-Mossotti factor. This is significant in that a particle may experience both dielectrophoresis and electrorotation simultaneously, and the magnitudes and directions of both are related to the interaction between the dielectric properties of particle and medium. As an example of this, a plot of the frequency-dependence of a solid particle exhibiting a single dielectric dispersion.

TRAVELLING-WAVE DIELECTROPHORESIS

Travelling-wave dielectrophoresis is a linear analogue of electrorotation; instead of electrodes being arranged in a circle, they are laid out as "tracks", but the relationship of phases (each electrode being 90° phase advanced from the last) remains. This produces an electric field wave which "travels" along the electrodes. When this wave interacts with a polarisable particle, a dipole is induced which moved with the electric field peak. If the travelling wave is moving sufficiently quickly, the dipole will lag behind the field in a similar manner to that observed in electrorotation. However, as the wave is travelling in a linear (rather than rotational) manner, the result is the induction of a force rather than a torque, with particles moving along the electrodes like a train travelling along tracks. This is demonstrated schematically in figure.

If the electric field E moves sufficiently quickly, the induced dipole M will lag behind the electric field, inducing a force F in the particle. This causes the particle to move along the electrodes, a phenomenon known as *travelling-wave dielectrophoresis. The value of the force* F_{TWD}*is given by equation*: where l is the wavelength of the travelling wave. This phenomenon was first observed by Batchelder. However, it remained largely unexplored until work by Masuda *et al.* on using *synchronously* induced motion in low frequency travelling fields that travelling wave dielectrophoresis became significant. This work led to Fuhr *et al.* effectively rediscovering asynchronous travelling wave dielectrophoresis. Since then, a large corpus of study has been conducted on this technique, including theoretical studies and demonstrations of practical devices for electrostatic pumping and biotechnological applications.

ELECTRODE CONFIGURATIONS

In practice, dielectrophoretic manipulation of particles is performed on planar electrode arrays. Such electrode arrays employ patterned electrodes fabricated of gold or a similar conductor deposited on glass or silicon. A common electrode configuration used in AC Electrokinetics research is a quadrupole arrangement where four electrodes point towards a central enclosed region, as shown in the schematic in figure.

The electrodes are typically fabricated of gold on a glass microscope slide. The gap between opposing electrodes in the centre of the array is typically of the order 10-50 μm across, but can be as small as 500nm or as large as 1mm.

To induce dielectrophoretic motion in particles suspended near the electrode array, electrodes would be energised such that a and c are of the same phase, and b and d are in antiphase to them. To cause electrorotation in particles within the central gap, electrodes b, c and d would be phase-shifted by 90°, 180° and 270° with respect to electrode a.

These electrodes may be of a variety of designs, according to the degree of field non-uniformity required. The advantage with this arrangement is twofold.

Firstly, electrodes can easily be used for dielectrophoresis, electrorotation or both. By energising the electrodes such that the phase angle of applied AC field between adjacent electrodes is 180°, particles will experience only dielectrophoresis. If the phase angle is reduced such that adjacent electrodes differ by 90°, the particle will experience both electrorotation and dielectrophoresis. The second advantage of the quadrupole design is that at the centre of the electrode array, there exists a well-defined electric field minimum, which is surrounded by well-defined field maxima at the electrode edges. This definition is advantageous when attempting to manipulate nanometre-sized particles, where Brownian motion is significant and may be sufficient to enable particles to "escape" from a weak forcefield "trap".

This is illustrated in figure, which shows a simulation of the electric field in a plane 5 μm above the plane of the electrodes shown in figure. The dark circular region at the centre of the electrode array is the field minimum in which particles become trapped by the surrounding ring of high field gradient. The field maximum extends along the inter-electrode gaps, where particles will accumulate by positive dielectrophoresis. The simulation was performed using a Method of Moments software suite developed by the author at the University of Wales at Bangor. As can be seen, there is an electric field minimum (dark region) at the centre of the electrode array, surrounded by a ring of high electric field gradient. Particles experiencing negative dielectrophoresis are repelled into this minimum and become trapped.

The electric field strength is high (white regions) along the electrode edges, where particles collect by positive dielectrophoresis.

Advances in fabrication technology such as the use of electron beam lithography have allowed the manufacture of electrodes with feature sizes of the order 100 nm, covering many square cm. By reducing the size of inter-electrode gaps, electric fields of 10^7 Vm^{-1} can be generated by applying 10V across a 1 μm inter-electrode gap. More advanced dielec-trophoretic traps have been developed in which two sets of electrodes were stacked so as to form an octopole field "cage". This type of geometry has allowed the controlled manipulation of populations of particles in all 3 dimensions.

3

Impact of Molecular Nanotechnology

Molecular nanotechnology is a speculative subfield of nanotechnology regarding the possibility of engineering molecular assemblers, machines which could re-order matter at a molecular or atomic scale. Regarding the risks from molecular manufacturing, an often cited worst-case scenario is "grey goo", a hypothetical substance into which the surface of the earth might be transformed by self-replicating nanobots running amok. This concept has been analysed by Freitas in "Some Limits to Global Ecophagy by Biovorous Nanoreplicators, with Public Policy Recommendations" With the advent of nan-biotech, a different scenario called green goo has been forwarded. Here, the malignant substance is not nanobots but rather self-replicating organisms engineered through nanotechnology.

The Center for Responsible Nanotechnology:

- Molecular manufacturing allows the cheap creation of incredibly powerful devices and products. How many of these products will we want? What environmental damage will they do? The range of possible damage is vast, from personal low-flying supersonic aircraft injuring large numbers of animals to collection of solar energy on a sufficiently large scale to modify the planet's albedo and directly affect the environment. Stronger materials will allow the creation of much larger machines, capable of excavating or otherwise destroying large areas of the planet at a greatly accelerated pace.

It is too early to tell whether there will be economic incentive to do this. However, given the large number of activities and purposes that would damage the environment if taken to extremes, and the ease of taking them to extremes with molecular manufacturing, it seems likely that this problem is worth worrying about. Some forms of damage can result from an aggregate of individual actions, each almost harmless by itself. Such damage is quite hard to prevent by persuasion, and laws frequently don't work either; centralized restriction on the technology itself may be a necessary part of the solution. Finally, the extreme compactness of nanomanufactured machinery will tempt the use of very small products, which can easily turn into nano-litter that will be hard to clean up and may cause health problems. The site list numerous other risks and benefits.

MOLECULAR NANOTECHNOLOGY

Modern surgery works only because cells have a remarkable ability to regroup, bury their dead and heal over the injury. Nanotechnology, "the manufacturing technology of the 21st century," should let us economically build a broad range of complex molecular machines (including, not incidentally, molecular computers). It will let us build fleets of computer controlled molecular tools much smaller than a human cell and built with the accuracy and precision of drug molecules. Such tools will let medicine, for the first time, intervene in a sophisticated and controlled way at the cellular and molecular level.

They could remove obstructions in the circulatory system, kill cancer cells, or take over the function of subcellular organelles. Just as today we have the artifical heart, so in the future we could have the artificial mitochondrion. Equally dramatic, nanotechnology will give us new instruments to examine tissue in unprecedented detail. Sensors smaller than a cell would give us an inside and exquisitely precise look at ongoing function.

Tissue that was either chemically fixed or flash frozen could be analysed literally down to the molecular level, giving a completely detailed "snapshot" of cellular, subcellular and molecular activities. There is broad agreement (though not consensus) that we will at some point in the future be able to inexpensively fabricate essentially any structure that is consistent with chemical and physical law and specified in molecular.

The most direct route to achieving this capability involves positioning and assembling individual atoms and molecules in a fashion conceptually similar to snapping together LEGO blocks. By designing and building programmable self replicating manufacturing systems that incorporate these principles we should be able to achieve very low manufacturing costs. While the design and development of such programmable self replicating molecular manufacturing systems will be a major task and will likely require many years or a few decades, it appears that this kind of capability, to quote Feynman. Design concepts for general purpose self replicating manufacturing systems have been discussed for many years, and their utility in manufacturing has been emphasized recently. These proposals draw on a body of work started by von Neumann.

A wide range of methods have been considered. The von Neumann architecture for a self replicating system is the ancestral and archetypal proposal.

VON NEUMANN ARCHITECTURE

Von Neumann's proposal consisted of two central elements: a universal computer and a universal constructor. The universal computer contains a programme that directs the behaviour of the universal constructor. The universal constructor, in turn, is used to manufacture both another universal computer and another universal constructor. Once construction is finished the programme contained in the original universal computer is copied to the new universal computer and programme execution is started.

Von Neumann worked out the details for a constructor that worked in a theoretical two-dimensional cellular automata world (parts of his proposal have since been modeled computationally). The constructor had an arm which it could move about and which could be used to change the state of the cell at the tip of the arm.

By progressively sweeping the arm back and forth and changing the state of the cell at the tip, it was possible to create "objects" consisting of regions of the two-dimensional cellular automata world which were fully specified by the programme that controlled the constructor. While this solution demonstrates the theoretical validity of the idea, von Neumann's kinematic constructor (which was not worked out in such detail) has had perhaps a greater influence, for it is a model of general manufacturing which can more easily be adapted to the three-dimensional world in which we live. The kinematic constructor was a robotic arm which moved in three-space and which grasped parts from a sea of parts around it. These parts were then assembled into another kinematic constructor and its associated control computer.

An important point to notice is that self replication, while important, is not by itself an objective. A device able to make copies of itself but unable to make anything else would not be very valuable. Von Neumann's proposals centered around the combination of a universal constructor, which could make anything it was directed to make, and a universal computer, which could compute anything it was directed to compute. It is this ability to make any one of a broad range of structures under flexible programmatic control that is of value. The ability of the device to make copies of itself is simply a means to achieve low cost, rather than an end in itself.

DREXLER'S ARCHITECTURE

Drexler's assembler follows the von Neumann kinematic architecture, but is specialized for dealing with systems made of atoms. The emphasis here (in contrast to von Neumann's proposal) is on small size. The computer and constructor both shrink to the molecular scale, while the constructor takes on additional detail consistent with the desire to manipulate molecular structures with atomic precision.

The tip chemistry is logically similar to the ability of the von Neumann universal constructor to alter the state of a cell at the tip of the arm, but now the change in "state" corresponds to a change in molecular structure.

That is, we must specify a set of well defined chemical reactions that take place at the tip of the arm, and this set must be sufficient to allow the synthesis of the structures of interest. It is worth noting that current methods in computational chemistry are sufficient to model the kinds of structures that will appear in a broad class of molecular machines, including all of the structures and reactions needed for some assemblers

SIZES

Even including system overhead (power, connections, etc). the volume per element should still be less than 100 cubic nanometers. A 10,000 element logic system (enough to hold a small processor) would occupy a cube no more than 100 nanometers on a side.

That is, a volume only slightly larger than 0.001 cubic microns would be sufficient to hold a small computer. This compares favourably with the volume of a typical cell (thousands of cubic microns) and is even substantially smaller than subcellular organelles.

Operating continuously at a gigahertz such a computer would use less than 10 ^ *-9* watts. By comparison, the human body uses about 100 watts at rest and more during exercise. Slower operation and the use of reversible logic would reduce power consumption, quite possibly dramatically. A variety of molecular sensors and actuators would also fit in such a volume. A molecular "robotic arm" less than 100 nanometers long should be quite feasible, as well as molecular binding sites 10 nanometers in size or less.

By contrast, a single red blood cell is about 8 microns in diameter (over 80 times larger in linear dimensions than our 100 nanometer processor). Devices of the size range suggested above (~0.1 microns) would easily fit in the circulatory system and would even be able to enter individual cells.

KILLING OF CANCER CELLS

Given such molecular tools, we could design a small device able to identify and kill cancer cells. The device would have a small computer, several binding sites to determine the concentration of specific molecules, and a supply of some poison which could be selectively released and was able to kill a cell identified as cancerous. The device would circulate freely throughout the body, and would periodically sample its environment by determining whether the binding sites were or were not occupied.

Occupancy statistics would allow determination of concentration. Today's monoclonal antibodies are able to bind to only a single type of protein or other antigen, and have not proven effective against most cancers.

The cancer killing device suggested here could incorporate a dozen different binding sites and so could monitor the concentrations of a dozen different types of molecules.

The computer could determine if the profile of concentrations fit a pre-programmed "cancerous" profile and would, when a cancerous profile was encountered, release the poison.

Beyond being able to determine the concentrations of different compounds, the cancer killer could also determine local pressure. A pressure sensor little more than 10 nanometers on a side would be sufficient to detect pressure changes of less than 0.1 atmospheres (a little over a pound per square inch. Linear increases in sensor volume would produce exponential increases in immunity to thermal noise and linear improvements in pressure sensitivity if that were to prove useful. Doubling the linear dimensions of the sensor would produce an eight-fold increase in both volume and pressure sensitivity). As acoustic signals in the megahertz range are commonly employed in diagnostics (ultrasound imaging of pregnant women, for example), the ability to detect such signals would permit the cancer killer to safely receive broadcast instructions. By using several macroscopic acoustic signal sources, the cancer killer could determine its location within the body much as a radio receiver on earth can use the transmissions from several satellites to determine its position (as in the widely used GPS system).

Megahertz transmission frequencies would also permit multiple samples of the pressure to be taken from the pressure sensor, as the CPU would be operating at gigahertz frequencies. The cancer killer could thus determine that it was located in (say) the big toe. If the objective was to kill a colon cancer, the cancer killer in the big toe would not release its poison. Very precise control over location of the cancer killer's activities could thus be achieved. The cancer killer could readily be reprogrammed to attack different targets (and could, in fact, be reprogrammed via acoustic signals transmitted while it was in the body). This general architecture could provide a flexible method of destroying unwanted structures (bacterial infestations, etc).

PROVIDING OXYGEN

A second application would be to provide metabolic support in the event of impaired circulation. Poor blood flow, caused by a variety of conditions, can result in serious tissue damage. A major cause of tissue damage is inadequate oxygen. A simple method of improving the levels of available oxygen despite reduced blood flow would be to provide an "artificial red blood cell." We will consider a simple design here: a sphere with an internal diameter of 0.1 microns (100 nanometers) filled with high pressure oxygen at ~1,000 atmospheres (about 10^8 pascals). The oxygen would be allowed to trickle out from the sphere at a constant rate (without feedback). Diamond has a Youngs modulus of about 10 ^ *12* pascals. An atomically precise diamondoid structure should be able to tolerate a stress of greater than 5×10^{10} pascals (5 per cent of the modulus).

Thus, a 0.1 micron sphere of oxygen at a pressure of 10 ^ *8* pascals could be contained by a hollow diamondoid sphere with an internal diameter of 0.1 microns and a thickness of less than one nanometer. This thickness, thin as it is, results in an applied stress on the diamond of well under 1 per cent of its modulus—from a purely structural point of view we should be able to use a very large "bucky ball," *i.e.*, a sphere whose surface is a single layer of graphite.

Perhaps the most complex issue involved in the selection of the material is the reaction of the body's immune system. While some suitable surface structure should exist which does not trigger a response by the immune system—after all, there are many surfaces in the body that are not attacked—the selection of a specific surface structure will require further research. To give a feeling for the range of possible surface structures, the hydrogenated diamond (111) surface could have a variety of "camouflauge" molecules covalently bound to its surface. A broad range of biological molecules could be anchored to the surface, either directly or via polymer tethers. The Van der Waals' equation of state is (p+a/v ^ 2) (v–b) = RT, where p is the pressure, v is the volume per mole, R is the universal gas constant, T is the temperature in Kelvins, and a and b are constants specific to the particular gas involved. For oxygen, a = 1.36 atm litre ^ *2*/mole ^ *2* and b = 0.03186 litre/mole and R = 0.0820568 litre-atmospheres/mole-kelvin. A mole of oxygen at 1,000 atmospheres and at body temperature (310 Kelvins) occupies 0.048 liters, or about 21 moles/litre. A mole of oxygen at 1 atmosphere and 310 Kelvins occupies 25.4 liters, or about 0.04 moles/litre.

This implies a compression of ~530 to 1. A resting human uses ~240 cc/minute of oxygen, so a litre of oxygen compressed to 1,000 atmospheres should be sufficient to maintain metabolism for about 36 hours (a day and a half). It might be desirable to replace less than a litre of blood with our microspheres of compressed oxygen, but it should still be quite feasible to provide oxygen to tissue even when circulation is severely compromised for periods of at least many hours from a single infusion.

Controlled release of oxygen from the diamondoid sphere could be done using the selective transport method proposed by Drexler and illustrated in figure. Transport in the "wrong" direction (for this application), but simply reversing the direction of rotor motion would result in transport from inside the reservoir to the external fluid.

More sophisticated systems would release oxygen only when the measured external partial pressure of oxygen fell below a threshold level, and so could be used as an emergency reserve that would come into play only when normal circulation was (for some reason) interupted.

Full replacement of red blood cells would involve the design of devices able to absorb and compress oxygen when the partial pressure was above a high threshold (as in the lungs) while releasing it when the partial pressure was below a lower threshold (as in tissues using oxygen). In this case, selective transport of oxygen into an internal reservoir would be required. If a single stage did not provide a sufficiently selective transport system, a multi-staged or cascaded system could be used.

Compression of oxygen would presumably require a power system, perhaps taking energy from the combustion of glucose and oxygen (thus permitting free operation in tissue). Release of the compressed oxygen should allow recovery of a significant fraction of the energy used to compress it, so the total power consumed by such a device need not be great. If the device were to simultaneously absorb carbon dioxide when it was present at high concentrations (in the tissue) and release it when it was at low concentrations (in the lungs), then it would also provide a method of removing one of the major products of metabolic activity.

Calculations similar to those given above imply a human's oxygen intake and carbon dioxide output could both be handled for a period of about a day by about a litre of small spheres. As oxygen is being absorbed by our artificial red blood cells in the lungs at the same time that carbon dioxide is being released, and oxygen is being released in the tissues when carbon dioxide is being absorbed, the energy needed to compress one gas can be provided by decompressing the other.

The power system need only make up for losses caused by inefficiencies in this process. These losses could presumably be made small, thus allowing our artificial red blood cells to operate with little energy consumption. By comparison, a litre of blood normally contains ~0.2 liters of oxygen, while one litre of our spheres contained ~530 liters of oxygen (where "litre of oxygen" means, as is common in the literature on human oxygen consumption, one litre of the gas under standard conditions of temperature and pressure).

Thus, our spheres are over 2,000 times more efficient per unit volume than blood; taking into account that blood is only about half occupied by red blood cells, our spheres are over 1,000 times more efficient than red blood cells. Failure of a 0.1 micron sphere would result in creation of a bubble of oxygen less than 1 micron in diameter. Occasional failures could be tolerated. Given the extremely low defect rates projected for nanotechnology, such failures should be very infrequent.

ARTIFICIAL MITOCHONDRIA

In particular, the mitochondria will, at some point, fail. Increased oxygen levels in the presence of non-functional or partially functional mitochondria will be ineffective in restoring the tissue. However, more direct metabolic support could be provided. The direct release of ATP, coupled with selective release or absorption of critical metabolites (using the kind of selective transport system mentioned earlier), should be effective in restoring cellular function even when mitochondrial function had been compromised.

The devices restoring metabolite levels, injected into the body, should be able to operate autonomously for many hours (depending on power requirements, the storage capacity of the device and the release and uptake rates required to maintain metabolite levels). While levels of critical metabolites could be restored, other damage caused during the ischemic event would also have to be dealt with. In particular, there might have been significant free radical damage to various molecular structures within the cell, including its DNA.

If damage was significant restoring metabolite levels would be insufficient, by itself, to restore the cell to a healthy state. Various options could be pursued at this point. If the cellular condition was deteriorating (unchecked by the normal homeostatic mechanisms, which presumably would cease to function when cellular energy levels fell below a critical value), some general method of slowing further deterioration would be desirable.

Cooling of the tissue, or the injection of compounds that would slow or block deteriorative reactions would be desirable. As autonomous molecular machines with externally provided power could be used to restore function, maintaining function in the tissue itself would no longer be critical. Deliberately turning off the metabolism of the cell to prevent further damage would become a feasible option.

Following some interval of reduced (or even absent) metabolic activity during which damage was repaired, tissue metabolism could be restarted again in a controlled fashion. It is clear that this approach should be able to reverse substantially greater damage than can be dealt with today. A primary reason for this is that autonomous molecular machines using externally provided power would be able to continue operating even when the tissue itself was no longer functional. We would finally have an ability to heal injured cells, instead of simply helping injured cells to heal themselves.

RESEARCH

Advances in medical technology necessarily depend on our understanding of living systems. With the kind of devices discussed earlier, we should be able to explore and analyse living systems in greater detail than ever before considered possible. Autonomous molecular machines, operating in the human body, could monitor levels of different compounds and store that information in internal memory.

They could determine both their location and the time. Thus, information could be gathered about changing conditions inside the body, and that information could be tied to both the location and the time of collection. Physical samples of small volumes (nano tissue samples) could likewise be taken. These molecular machines could then be filtered out of the blood supply and the stored information (and samples) could be analysed. This would provide a picture of activities within healthy or injured tissue. This new knowledge would give us new insights and new approaches to curing the sick and healing the injured.

More dramatically, it should be feasible to take "snapshots" of tissue samples and analyse the structure down to the molecular level. First, a small tissue sample could be either fixed or frozen. Chemical fixation can be used to rapidly block most tissue changes. Ultra fast freezing of small tissue samples is an effective method of halting essentially all chemical processes and diffusion of all molecules. Once fixed or frozen, the tissue sample could be analysed in a leisurely fashion. With nanotechnology (and indeed, to some extent with current STM and AFM technologies, though rather more expensively) it should be feasible to scan the tissue surface in molecular detail, and store that information in a computer. Once the surface had been scanned, it could be removed in a very selective and precise fashion, and scanned again.

As an example, the use of a positionally controlled carbene has been proposed for use in the synthesis of complex diamondoid structures. Such a positionally controlled carbene is highly reactive and, if positioned at an appropriate site on the surface of the tissue being analysed, would readily react with a surface molecule. This surface molecule could then be removed. A wide variety of other "sticky" molecular tools could be brought up to the surface and allowed to react with surface molecules, which could then be removed, exposing the layers beneath. The use of a positionally controlled carbene implies that the environment in which it is used must be inert.

This requirement could be satisfied by analysing the tissue sample at very low temperature (a few Kelvins) and in a very good vacuum. Under these conditions the tissue specimen would remain stable during even a protracted analysis process. While this process can readily be envisioned for very small structures, nanotechnology should make massive parallelism feasible. That is, a single positional device could be used at a certain speed to provide information about a certain (rather small) volume of tissue in a reasonable time.

Nanotechnology should permit the manufacture of a large number of small devices, each able to analyse a small volume. Given enough such devices operating in parallel, larger volumes could be analysed and the information from many individual devices integrated to provide a coherent picture of the larger whole. Effective use of this option will require massive computational power—which will also be made feasible with nanotechnology. Estimates of the computational power that should be provided by nanotechnology exceed 10 ^ *24* logic operations per second for a single desktop computer.

This amount of raw computational power should make control of a large number of parallel devices feasible, and should permit integration and analysis of the information so obtained.

In short, tissue samples could be "frozen" (either literally by ultrafast cooling or figuratively by chemical fixation) and the entire resulting tissue sample could be analysed down to the level of individual molecules. The information so obtained could be processed by computers able to handle the flood of data produced. The resulting "snapshots" will provide us with an instantaneous look at metabolic and cellular activities across even relatively large volumes of tissue. Such an ability should revolutionize our understanding of the complex processes that take place in living systems.

The possibility of truly revolutionary advances in our medical abilities has also created renewed interest in cryonics. The abilities discussed here might well take years or decades to develop. It is quite natural to ask: "When might we see these systems actually used?" The scientifically correct answer is, of course, "We don't know." That said, it is worth noting that if progress in computer hardware continues as the trend lines of the last 50 years suggest, we should have some form of molecular manufacturing in the 2010 to 2020 time frame. After this, the medical applications will require some additional time to develop. The remarkably steady trend lines in computer hardware, however, give a false sense that there is a "schedule" and that developments will spontaneously happen at their appointed time.

CONTROL TO CHEMICAL SYNTHESIS

Manufactured products are made from atoms. The properties of those products depend on how those atoms are arranged. Viewed from the molecular level today's macroscopic manufacturing methods are crude and imprecise. Casting, milling, welding and all the other traditional manufacturing methods spray atoms about in great statistical herds. Even lithography (which already lets us put millions of transistors on a chip no bigger than your fingernail) is fundamentally statistical and random.

Exactly how many dopant atoms are in a single transistor and exactly where each individual dopant atom is located is neither specified nor known: if we have roughly the right number in roughly the right place, we can make a working transistor. For today, that is good enough.

The exception is chemistry. Large high purity crystals can have almost every atom in the right place. So, too, can many long polymers. The structures of proteins with hundreds and even thousands of amino acids can be specified down to the last atom. Most dramatically (and fortunately for us!) DNA strands with many tens of millions of bases can be copied with almost perfect accuracy. And it seems that almost any small molecule (with perhaps several dozens of atoms) can be synthesized, if only we have the skill and patience. Yet the laws of physics and chemistry in principle permit arranging and rearranging the elements in so many combinations and permutations that all of our manufacturing skills and all of our chemical skills barely suffice to scratch the surface of what is possible.

THE UTILITY OF DIAMOND

Almost any manufactured product could be improved, often by several orders of magnitude, if we could precisely control its structure at the molecular level. We often want our products to be light and strong. Diamond is light and strong: the strength-to-weight ratio of diamond is over 50 times that of steel. Yet we do not today have diamond spars in airplanes nor diamond hulls for rockets. Today we can't economically make diamond. Even if we could, simple diamond crystals can shatter. We'd have to modify the structure to make it tough and shatter proof: perhaps diamond fibres. While easily done in principle, we can't do this in practice today. Great strength and light weight are not the exclusive province of diamond: graphite can be stronger. And if we consider the many ways in which carbon atoms can be arranged and rearranged, then it's obvious that there are a host of other possibilities.

Yet all share a common problem: we can't yet economically make them in the exact shapes that we want. Great strength is only one property that we prize highly: when we make computers we are more concerned by electrical properties. Here, too, diamond excels.

Today's computers are made of semiconductors, and the semiconductor of choice is silicon. This is not because silicon is the ideal semiconductor from which to make computers, but because we know how to make devices from it. The computer industry has strong opinions about what makes a good logic device and what makes a good computer and diamond will let us make better computers than silicon. Diamond has a wider bandgap, hence electrical devices will work at higher temperatures. It has greater thermal conductivity, so devices can be more easily cooled. It has a greater breakdown field, hence devices can be smaller. It has higher electron and hole mobility which, when combined with higher electric fields, will result in higher speed.

But again, we see no diamond computers, just as we see no diamond airplanes: we can't economically manufacture them yet. Large pure crystals of silicon can be made relatively easily, but large pure crystals of diamond are scarce. We can etch the silicon surface and add dopants with a precision measured in tenths of microns, while the corresponding steps for diamond are more difficult. Not more difficult in principle: just more difficult today. Making computers highlights another problem. It's not enough to make a pure crystal, it must also have an extremely precise and complex pattern of impurities. The exact location of the dopant atoms in the semiconductor lattice controls how devices function and where signals can propagate.

Local order is crucial to make each device work, but long range complex order is crucial to make the computer as a whole work.

While we can make some things today that are highly precise and have simple long range order (*e.g.*, crystals), it is the requirement for complex long range order that prevents us from making computers of the kind we'd like to make.

While it's plausible we could make high density memory from crystals and perhaps some types of cellular automatons, we couldn't make anything that resembled the computers on the market today.

Today's high speed semiconductor-based digital computers (like the 80486 or the Pentium) have millions of logic elements wired together in complex and highly idiosyncratic patterns. This is well beyond the capabilities of crystal growth or bio-polymer synthesis. It will require a fundamentally new manufacturing technology: molecular manufacturing.

Today, there is a gap in our synthetic abilities: we can make complex mechanical machinery and electronic devices (including computers, which have millions of transistors), but we can't make such devices with the precision with which the chemist can synthesize a crystal, a bio-polymer, or a relatively small molecule. With chemistry we can make precise molecular structures and compounds, but we haven't been able to scale up that success to molecular computers (and other macroscopic products as precise as molecules).

Molecular manufacturing will, by definition, let us economically manufacture almost any specified structure that is consistent with the laws of chemistry and physics. To simplify the problem somewhat we can narrow our focus to structures that resemble diamond in a broad sense: the diamondoid structures as defined by Drexler. This class includes (among other things) diamond crystals of arbitrary shape but with stably terminated surfaces (*e.g.*, hydrogenated (111) or the like) and with impurities at precise locations in the diamond lattice (*e.g.*, substitutional boron). Our objective is to manufacture particular diamondoid structures once the location and type of every atom has been specified by design.

COMPUTER INDUSTRY

The attraction of molecular manufacturing for the computer industry should be clear. It should let us make computers at a manufacturing cost of less than a dollar per pound, operating at frequencies of tens of gigahertz or more, with linear dimensions for a single device of roughly 10 nanometers, high reliability, and energy dissipation (using conventional methods) of roughly 10 ^ –18 joules per logic operation. If we make thermodynamically reversible computers (which the author and others have recently shown can be made from conventional electronic devices, *e.g.*, CMOS) then the energy dissipation per logic operation can be reduced to well below kT at T = 300 Kelvins (well below 10 ^ –21 joules).

The computer industry is spending billions of dollars to make better computers. It is widely acknowledged within the industry that lithography is approaching its limits.

Articles like *The Future of the Transistor*, *Miniaturization of Electronics and its Limits* and *Outlook for VLSI: Will the Balloon Burst?* Quite clearly show that conventional lithography will run out of steam (in perhaps a decade, though there is less agreement about the exact time frame). There is already interest in molecular logic devices and that interest will increase sharply as improvements in conventional manufacturing methods become increasingly difficult. However, any new proposal for manufacturing molecular computers will be weighed against (at least) the criteria mentioned above.

If it cannot easily beat conventional methods after they have been pushed to their uttermost limits, then it will be rejected. The computer industry will soon be pouring vast sums into research aimed at molecular computing, but the great bulk of funding will go towards well thought out proposals that offer a realistic possibility of substantially exceeding the performance of the ultimately evolved silicon VLSI technology that we expect to develop over the next decade. If you can't beat tomorrow's mainstream computers, you might as well not try. For this and many other reasons the class of diamondoid structures is a reasonable one to consider.

The problem of building a diamondoid electronic computer captures many of the fundamental issues in molecular manufacturing, and poses clearly the issue of building large structures that cannot be made by regular repetition of some substructure (*e.g.*, the unit cell of a crystal or the monomeric unit in a bio-polymer).

This brings us to a core issue in molecular manufacturing: how do we synthesize such things? Today, we can synthesize diamond at low pressure and low temperature by using CVD (Chemical Vapour Deposition) methods.

Diamond CVD growth involves highly reactive species (radicals, carbenes, etc.) in a gas over the growing diamond surface that bombard and react with that surface at random. Because reaction sites are random, growth of many defect structures occurs (dislocations, etc.) as well as the desired perfect diamond structure. Two fundamental mechanisms in the growth process include:

- Abstraction of hydrogens from the diamond surface leaving behind reactive sites (dangling bonds, radicals) and
- Interaction of carbon species (both reactive (CH2, CH3, etc.) as well as relatively unreactive species (C2H2)) with the surface, thus depositing carbon.

If we are to synthesize diamondoid structures it is plausible that we begin our search for the basic reaction steps involved in this synthesis by looking at existing reactions that occur in the CVD growth of diamond. The use of a reactive gas in the synthesis process, however, would seem to defeat any hope of making precisely patterned diamondoid structures, for the gas will interact with the growing surface at random.

POSITIONAL CONTROL IS FUNDAMENTAL

Here, we introduce the fundamental concept of molecular manufacturing: positional control over the site of reactions. To take a specific example we consider site specific hydrogen abstraction from the diamond (111) surface.

The ability to remove specific hydrogen atoms from the surface of the diamondoid work piece under construction is likely to be a fundamental unit operation in any attempt to make atomically precise diamondoid structures.

Hydrogen abstraction during CVD diamond growth typically involves a radical reaction between atomic H from the gas with H bonded to carbon on the surface producing H2. It is unclear how to make this process site specific. However, there are other structures with a high affinity for hydrogen which offer greater possibilities for positional control. In particular, the propynyl radical C3H3 has a great affinity for hydrogen.

Further, this radical has the very useful property that it has two ends: one end is a highly reactive radical while the other end is a stable sp3 carbon. Thus, we could synthesize a larger molecule with the propynyl radical at its end. The larger molecule would be held at the tip of a positional device.

The positional device would provide control over the orientation and position of this hydrogen abstraction tool (*e.g.*, a six degrees of freedom manipulator) and thus control the site of abstraction by controlling the position of the tool. Ab initio quantum chemical analysis of the abstraction of hydrogen from isobutane using an ethynyl abstraction tool supports the idea that the barrier to this reaction is zero. The reaction will proceed rapidly and, because of the large exothermicity, irreversibly.

Calculated barriers for abstraction from several other molecules were also small, suggesting that this hydrogen abstraction tool could be used to abstract hydrogen from a wide range of different molecules. The site specific abstraction of hydrogen illustrates the core concept in molecular manufacturing: selecting the reaction site by controlling the position and orientation of the reactants. The (relatively stiff) diamondoid workpiece is held in place, while a tool (in our example, a hydrogen abstraction tool) is positioned using a rather conventional (if also rather small) robotic arm. The ideas of using tools, controlling the position of those tools with a general purpose manipulator, and building complex structures by putting together components using those positionally controlled tools are rather common and even mundane at the macroscopic level.

At the molecular level, they are new and almost shocking: yet it is simply mapping onto the molecular world the concepts and ideas that have proven so useful and powerful in macroscopic manufacturing. By adding positional control we should be able to develop a method of molecular manufacturing which combines the best features of both conventional macroscopic manufacturing and chemical synthesis.

MOLECULAR TOOLS

If we are to grow diamond, we must also have carbon deposition tools. Drexler has suggested the use of positionally controlled carbenes and alkynes and proposed reaction pathways and surface structures where these tools would apply. In both cases, the tools are positioned at a precise point on the growing diamondoid structure and are used to deposit one or more carbon atoms at a desired location.

These deposition reactions parallel proposals in the CVD literature except for the addition of positional control (*e.g.*, at least one portion of the moiety must be part of an extended "handle" which can be held by a positional device). The broad range of possible tools coupled with the great power of ab initio computational chemistry should let us define and verify a complete set of molecular tools capable of synthesizing essentially any diamondoid structure.

TOOL USE

For such tools to be usable in a system context we must satisfy certain constraints. First and foremost, we must have a device capable of positioning the tool to within something like an atomic diameter. On the diamond (111) surface, the distance between adjacent hydrogens is about 2.5 Angstroms. Thus, positional accuracy of 1 to 2 angstroms for the hydrogen abstraction tool is required to prevent abstraction of the wrong hydrogen. Second, because the tools can be highly reactive, we require an inert environment. A simple inert environment is vacuum.

Compressed helium or some other inert gas would also work. Third, because it is the relative tool-workpiece position that must be controlled, the workpiece under construction must be relatively rigid (*e.g.*, not subject to vibrational motions that would exceed about an angstrom). Fourth and last, we must have some way of generating the sometimes highly reactive tools (*e.g.*, we need to define a precursor to the hydrogen abstraction tool, as well as precursors to the other tools).

In some sense, the analysis that we will now pursue is similar in type to retrosynthetic analysis. We start with the final product that we wish to build (a macroscopic diamondoid computer, for example), and then consider the possible predecessor structures which would yield the final product in one step. Then we consider the predecessor structures to those structures, and so on. We extend retrosynthetic analysis beyond its traditional bounds, but the general concept remains the same: given the finished product we deduce the possible ways in which it could have been constructed. This approach has been called "backwards chaining" by Drexler.

Work with SPMs (Scanning Probe Microscopes) clearly show that it is possible to achieve positional accuracies of a small fraction of an angstrom. Small (~0.1 microns) diamondoid "arms" or positional devices with similar positional accuracy are in principle quite feasible. The field of robotics provides a broad range of designs for positional devices which are largely scale independent. Shrinking these designs to submicron size is conceptually straightforwards. A factor of crucial importance in the design of molecular-scale positional devices is the accuracy with which the tip can be positioned, particularly in the face of thermal noise.

While atomically precise bearings and joints will not suffer from chatter, backlash, wear, tooth-to-tooth errors and other sources of inaccuracy caused by imprecise manufacturing, they will still suffer from positional errors caused by thermal noise. To control this source of error, it is essential that the robotic arm be very stiff, and so the use of stiff materials is desirable.

The Young's modulus for diamond is about 10 ^ 12 Pascals (very stiff), and back-of-the-envelope calculations show that a hollow cylinder of such material that is perhaps 100 nanometers long and 30 nanometers in diameter should have a positional accuracy at the tip, in the face of thermal noise at room temperature, of a small fraction of an atomic diameter.

A more detailed design and analysis of a jointed tubular robotic arm taking into account the bending and rocking motions of joints in the arm further supports this conclusion. Alternatives to the simple robotic arm are available which might be more attractive.

Other Requirements for Tool Use

Creating an inert environment also presents no fundamental problems: high quality vacuums are common in laboratories today. If our objective is to have a very small very high quality vacuum, then a relatively thin wall of diamondoid material could be used as a barrier to keep a volume which was a modest fraction of a cubic micron free of any contaminants. If the volume were initially constructed free of contaminants then such a barrier would keep the inside free of any contaminants with high probability.

Because we are building diamondoid structures, they will be very stiff. As a consequence, it is relatively easy to meet the requirement that the objects that are being manufactured must themselves be stiff. Finally, generation of "activated" tools from relatively stable precursors can be done by a variety of methods. Because we are assuming an environment in which we have positional control we can use particularly simple precursors.

TRANSPORT ACROSS A BARRIER

Having introduced a diamondoid barrier to keep unwanted contaminants out (much as the bacterial wall allows bacteria to maintain an appropriate internal environment in the face of a fluctuating external environment) we must now solve the problem of getting desired raw materials through the barrier. We might, for example, wish to transport the hydrogen abstraction tool precursor across the diamondoid barrier. After use we will also need to eject the spent tool. Several ways to solve this problem are feasible. A proposal by Drexler is to use a rotor embedded in the diamondoid wall which moves binding sites from the outside of the wall to the inside of the wall.

By modulating the affinity of the binding site so that it will have high affinity for the desired molecule outside the barrier and low affinity inside the barrier, efficient transport across the barrier can be achieved. The desired molecule will bind to the binding site when it is outside the barrier, the binding site will be rotated to the inside of the barrier and the binding affinity reduced (in the illustrated proposal by mechanically pushing a rod into the binding site, thus physically precluding occupancy), and the molecule will be released on the inside of the barrier. The result is to increase the concentration of the desired molecule. A few stages of such a filtration system can achieve extremely high purities. The final stage, rather than ejecting the molecule into a liquid, would deliver the molecule into the inert internal environment in a well defined orientation where it could be further processed.

CONTROL SIGNALS

Finally, we will need a source of control signals for our molecular arm. One general approach would be to use a molecular computer. We will not consider a particular design for a molecular computer here, it is sufficient to note that many proposals for molecular computation have been considered in the literature and it is generally expected that some type of very small computational device will be feasible in a few decades. This completes our (all too brief) outline of a small device able to manufacture a broad class of diamondoid materials. Basically, the design is driven by the desire to provide the environment needed to synthesize diamond and diamondoid materials using the kinds of reactions that occur naturally during CVD growth of diamond. The device is itself made from diamondoid materials, which means that one such device can manufacture a second such device.

This ability to self replicate is crucial in achieving low manufacturing costs. As the reader might appreciate, the design and construction of one such general purpose device might well prove to be a time consuming and expensive undertaking. This cost cannot be justified unless the resulting device has great value.

If the device can self replicate then the successful design and development of one such device can be used to build an entirely new manufacturing technology. The manufacturing costs for the second, third, fourth.... 10 ^ 10.... etc. devices will consist largely of the raw materials and energy costs. Thus, a very large R&D cost can (if necessary) be justified.

The major driving force in the previous design was the desire to approach fundamental limits (strength, stiffness, thermal conductivity, electrical characteristics, etc.) in the manufactured products. This in turn implied the use of diamondoid materials, which, when coupled with the known chemistry of CVD diamond synthesis resulted in a high vacuum system with highly reactive intermediates.

Relaxing the materials requirements gives us a *much* wider range of possible structures. In particular, we can consider what is sometimes called "brick" or "building block" based nanotechnology. In this approach we first design a set of molecular building blocks, and then assemble the building blocks by the use of positional control.

The ribosome can be viewed as the prototype for this approach. The building blocks are amino acids, and they are linked together by the ribosome to form proteins.

Our approach differs in two principle respects: first, we add positional and orientational control over the building blocks in three dimensions, while the ribosome can only build structures that are fundamentally one dimensional (relying on linear structures that spontaneously fold into a particular shape to achieve a degree of control in three dimensions).

Second, rather than using relatively floppy polymers, we prefer relatively rigid bricks that can be bonded to each other in a stiff three-dimensional framework. This greatly simplifies the system. Indeed, with brick-based nanotechnology one can relatively easily envision the synthesis of a set of bricks that can, with the addition of positional control, be assembled into a wide range of structures with the stiffness of (say) wood.

A conceptually simple and relatively near-term way to achieve some degree of positional control would be to use conventional SPMs. The tips used in current SPM's are usually quite crude, and even when it is possible to make a very fine tip the range of possible structures is very limited (tungsten or some other simple material is generally all that's available). The design of tips for the SPM that incorporate individual molecules specifically synthesized for the purpose is a likely next step, and one that seems essential if we are to make progress in using SPM's to guide chemical reactions in a selective way. Such a "molecular manipulator" should be within reach of today's experimental technology. While the Young's modulus of the things it could make would be substantially inferior to that of diamond, this can be compensated by making them bigger. Scaling laws are such that increasing the size of an object by a factor of 10 also increases its stiffness by a factor of 10, and so reduces the positional inaccuracy from thermal vibration by a factor of 10.

If we wish to build a molecular "arm" out of bricks, then to achieve the same positional accuracy as with a diamondoid arm we will have to make the arm bigger—perhaps several tenths of a micron or more—but the basic design concepts that were discussed previously for use with a diamondoid arm still hold. We can still position the individual bricks accurately both in position and orientation, we can still build larger structures by putting together many bricks, we can still control the synthesis process by positional means, we can still make a positional device from bricks, etc.

Why pursue such an approach when it can only make relatively inferior materials?

There are three primary reasons:

1. It's easier to do
2. It could still make many things that are very valuable by today's standards and

3. Such systems could be used to make better systems (*e.g.*, diamondoid systems).

As the reader will appreciate, there are many possible candidates for bricks. Many researchers are already considering the design of molecular building blocks, although in most cases they do not consider positional control. Krummenacker discusses several of the issues surrounding the design of molecular bricks intended for assembly via positional control.

Adding positional control makes the synthesis of a broad range of molecular structures feasible, but at the same time requires the design and synthesis of a mechanism able to provide such control.

The obvious first step is to provide positional control using relatively modest extensions to today's SPMs. The most significant addition is the inclusion of a molecular tip, for today's SPMs typically have a limited range of possible tip structures and are often imprecise at the molecular scale.

HEALTH AND SAFETY IMPACT FROM NANOPARTICLES

The presence of nanomaterials is not in itself a threat. It is only certain aspects that can make them risky, in particular their mobility and their increased reactivity. Only if certain properties of certain nanoparticles were harmful to living beings or the environment would we be faced with a genuine hazard. In this case it can be called nanopollution.

In addressing the health and environmental impact of nanomaterials we need to differentiate between two types of nanostructures:

- Nanocomposites, nanostructured surfaces and nanocomponents, where nanoscale particles are incorporated into a substance, material or device; and
- "Free" nanoparticles, where at some stage in production or use individual nanoparticles of a substance are present.

These free nanoparticles could be nanoscale species of elements, or simple compounds, but also complex compounds where for instance a nanoparticle of a particular element is coated with another substance. There seems to be consensus that, although one should be aware of materials containing fixed nanoparticles, the immediate concern is with free nanoparticles.

Nanoparticles are very different from their everyday counterparts, so their adverse effects cannot be derived from the known toxicity of the macro-sized material. This poses significant issues for addressing the health and environmental impact of free nanoparticles. To complicate things further, in talking about nanoparticles it is important that a powder or liquid containing nanoparticles almost never be monodisperse but contain instead a range of particle sizes. This complicates the experimental analysis as larger nanoparticles might have different properties from smaller ones. Also, nanoparticles show a tendency to aggregate, and such aggregates often behave differently from individual nanoparticles.

The lethal dose over six months for lab rats, of different kinds of nanoparticles are often characterized by a Skov Kjaer index, named after the scientist Kasper Skov Kjaer. The National Institute for Occupational Safety and Health has conducted initial research on how nanoparticles interact with the body's systems and how workers might be exposed to nano-sized particles in the manufacturing or industrial use of nanomaterials. NIOSH currently offers interim guidelines for working with nanomaterials consistent with the best scientific knowledge.

In "The Consumer Product Safety Commission and Nanotechnology," E. Marla Felcher suggests that the Consumer Product Safety Commission, which is charged with protecting the public against unreasonable risks of injury or death associated with consumer products, is ill-equipped to oversee the safety ofcomplex, high-tech products made using nanotechnology. Longer-term concerns center on the impact that new technologies will have for society at large, and whether these could possibly lead to either a post-scarcity economy, or alternatively exacerbate the wealth gap between developed and developing nations. The effects of nanotechnology on the society as a whole, on human health and the environment, on trade, on security, on food systems and even on the definition of "human", have not been characterized or politicized.

HEALTH ISSUES

The health impact of nanotechnology are the possible effects that the use of nanotechnological materials and devices will have on human health. As nanotechnology is an emerging field, there is great debate regarding to what extent nanotechnology will benefit or pose risks for human health. Nanotechnology's health impact can be split into two aspects: the potential for nanotechnological innovations to have medical applications to cure disease, and the potential health hazards posed by exposure to nanomaterials. Nanotoxicology is the field which studies potential health risks of nanomaterials.

The extremely small size of nanomaterials means that they are much more readily taken up by the human body than larger sized particles. How these nanoparticles behave inside the organism is one of the significant issues that needs to be resolved.

The behaviour of nanoparticles is a function of their size, shape and surface reactivity with the surrounding tissue. Apart from what happens if non-degradable or slowly degradable nanoparticles accumulate in organs, another concern is their potential interaction with biological processes inside the body: because of their large surface, nanoparticles on exposure to tissue and fluids will immediately adsorb onto their surface some of the macromolecules they encounter.

The large number of variables influencing toxicity means that it is difficult to generalise about health risks associated with exposure to nanomaterials–each new nanomaterial must be assessed individually and all material properties must be taken into account. Health and environmental issues combine in the workplace of companies engaged in producing or using nanomaterials and in the laboratories engaged in nanoscience and nanotechnology research. It is safe to say that current workplace exposure standards for dusts cannot be applied directly to nanoparticle dusts. Nanomedicine is the medical application of nanotechnology.

The approaches to nanomedicine range from the medical use of nanomaterials, to nanoelectronic biosensors, and even possible future applications of molecular nanotechnology. Nanomedicine seeks to deliver a valuable set of research tools and clinically helpful devices in the near future. The National Nanotechnology Initiative expects new commercial applications in the pharmaceutical industry that may include advanced drug delivery systems, new therapies, and in vivo imaging. Neuro-electronic interfaces and other nanoelectronics-based sensors are another active goal of research. Further down the line, the speculative field of molecular nanotechnology believes that cell repair machines could revolutionize medicine and the medical field.

ENVIRONMENTAL ISSUES

Nanopollution is a generic name for all waste generated by nanodevices or during the nanomaterials manufacturing process. This kind of waste may be very dangerous because of its size. It can float in the air and might easily penetrate animal and plant cells causing unknown effects. Most human-made nanoparticles do not appear in nature, so living organisms may not have appropriate means to deal with nanowaste. To properly assess the health hazards of engineered nanoparticles the whole life cycle of these particles needs to be evaluated, including their fabrication, storage and distribution, application and potential abuse, and disposal. The impact on humans or the environment may vary at different stages of the life cycle.

Environmental assessment is justified as nanoparticles present novel environmental impacts. Scrinis raises concerns about nano-pollution, and argues that it is not currently possible to "precisely predict or control the ecological impacts of the release of these nano-products into the environment." On the other hand, some possible future applications of nanotechnology have the potential to benefit the environment. Nanofiltration, based on the use of membranes with extremely small pores smaller than 10 nm are suitable for a mechanical filtration for the removal of ions or the separation of different fluids.

Furthermore, magnetic nanoparticles offer an effective and reliable method to remove heavy metal contaminants from waste water. Using nanoscale particles increases the efficiency to absorb the contaminants and is comparatively inexpensive compared to traditional precipitation and filtration methods. Furthermore, nanotechnology could potentially have a great impact on clean energy production. Research is underway to use nanomaterials for purposes including more efficient solar cells, practical fuel cells, and environmentally friendly batteries.

A NEED FOR REGULATION

Significant debate exists relating to the question of whether nanotechnology or nanotechnology-based products merit special government regulation. This debate is related to the circumstances in which it is necessary and appropriate to assess new substances prior to their release into the market, community and environment. Regulatory bodies such as the United States Environmental Protection Agency and the Food and Drug Administration in the U.S. or the Health and Consumer Protection Directorate of the European Commission have started dealing with the potential risks posed by nanoparticles. So far, neither engineered nanoparticles nor the products and materials that contain them are subject to any special regulation regarding production, handling or labelling. The Material Safety Data Sheet that must be issued for some materials often does not differentiate between bulk and nanoscale size of the material in question and even when it does these MSDS are advisory only.

Limited nanotechnology labeling and regulation may exacerbate potential human and environmental health and safety issues associated with nanotechnology. It has been argued that the development of comprehensive regulation of nanotechnology will be vital to ensure that the potential risks associated with the research and commercial application of nanotechnology do not overshadow its potential benefits. Regulation may also be required to meet community expectations about responsible development of nanotechnology, as well as ensuring that public interests are included in shaping the development of nanotechnology.

SOCIETAL IMPACT

Beyond the toxicity risks to human health and the environment which are associated with first-generation nanomaterials, nanotechnology has broader societal impact and poses broader social challenges. Social scientists have suggested that nanotechnology's social issues should be understood and assessed not simply as "downstream" risks or impacts. Rather, the challenges should be factored into "upstream" research and decision making in order to ensure technology development that meets social objectives.

Many social scientists and organizations in civil society suggest that technology assessment and governance should also involve public participation Societal risks from the use of nanotechnology have also been raised. On the instrumental level, these include the possibility of military applications of nanotechnology as well as enhanced surveillance capabilities through nano-sensors. The last few years has seen a gold rush to claim patents at the nanoscale. Over 800 nano-related patents were granted in 2003, and the numbers are increasing year to year.

Corporations are already taking out broad-ranging patents on nanoscale discoveries and inventions. For example, two corporations, NEC and IBM, hold the basic patents on carbon nanotubes, one of the current cornerstones of nanotechnology.

Carbon nanotubes have a wide range of uses, and look set to become crucial to several industries from electronics and computers, to strengthened materials to drug delivery and diagnostics. Carbon nanotubes are poised to become a major traded commodity with the potential to replace major conventional raw materials. However, as their use expands, anyone seeking to manufacture or sell carbon nanotubes, no matter what the application, must first buy a license from NEC or IBM.

POTENTIAL BENEFITS AND RISKS FOR DEVELOPING COUNTRIES

Nanotechnologies may provide new solutions for the millions of people in developing countries who lack access to basic services, such as safe water, reliable energy, health care, and education.

The United Nations has set Millennium Development Goals for meeting these needs. The 2004 UN Task Force on Science, Technology and Innovation noted that some of the advantages of nanotechnology include production using little labour, land, or maintenance, high productivity, low cost, and modest requirements for materials and energy.

Potential opportunities of nanotechnologies to help address critical international development priorities include improved water purification systems, energy systems, medicine and pharmaceuticals, food production and nutrition, and information and communications technologies. Nano-techno-logies are already incorporated in products that are on the market. Other nanotechnologies are still in the research phase, while others are concepts that are years or decades away from development.

Protection of the environ-ment, human health and worker safety in developing countries often suffers from a combination of factors that can include but are not limited to lack of robust environmental, human health, and worker safety regulations; poorly or unenforced regulation which is linked to a lack of physical and human capacity.

Often, these nations require assistance, particularly financial assistance, to develop the scientific and institutional capacity to adequately assess and manage risks, including the necessary infrastructure such as laboratories and technology for detection. However, concerns are frequently raised that the claimed benefits of nanotechnology will not be evenly distributed, and that any benefits associated with nanotechnology will only reach affluent nations. The majority of nanotechnology research and development-and patents for nanomaterials and products-is concentrated in developed countries.

In addition, most patents related to nanotechnology are concentrated amongst few multinational corporations, including IBM, Micron Technologies, Advanced Micro Devices and Intel.

This has led to fears that it will be unlikely that developing countries will have access to the infrastructure, funding and human resources required to support nano-technology research and development, and that this is likely to exacerbate such inequalities.

Producers in developing countries could also be disadvan-taged by the replacement of natural products by developments in nanotech-nology.

These natural products are important export crops for developing countries, and many farmers' livelihoods depend on them. It has been argued that their substitution with industrial nano-products could negatively affect the economies of developing countries, that have traditionally relied on these export crops.

EFFECTS ON LABORERS

Ray Kurzweil has speculated in *The Singularity is Near* that people who work in unskilled labour jobs for a livelihood may become the first human workers to be displaced by the constant use of nanotechnology in the workplace, noting that layoffs often affect the jobs based around the lowest technology level before attacking jobs with the highest technology level possible. It has been noted that every major economic era has stimulated a global revolution both in the kinds of jobs that are available to people and the kind of training they need to achieve these jobs, and there is concern that the world's educational systems have lagged behind in preparing students for the "Nanotech Age". It has also been speculated that nanotechnology may give rise to nanofactories which may have superior capabilities to conventional factories due to their small carbon and physical footprint on the global and regional environment.

The miniaturization and transformation of the multi-acre conventional factory into the nanofactory may not interfere with their ability to deliver a high quality product; the product may be of even greater quality due to the lack of human errors in the production stages.

Nanofactory systems may use precise atomic precisioning and contribute to making superior quality products that the "bulk chemistry" method used in 20th century and early 21st currently cannot produce. These advances might shift the computerized workforce in an even more complex direction, requiring skills in genetics, nanotechnology, and robotics.

SOCIETAL IMPACT OF NANOTECHNOLOGY

The societal impact of nanotechnology are the potential benefits and challenges that the introduction of novel nanotechnological devices and materials may hold for society and human interaction. The term is sometimes expanded to also include nanotechnology's health and environmental impact, but this article will only consider the social and political impact of nanotechnology. As nanotechnology is an emerging field and most of its applications are still speculative, there is much debate about what positive and negative effects that nanotechnology might have.

Beyond the toxicity risks to human health and the environment which are associated with first-generation nanomaterials, nanotechnology has broader societal implications and poses broader social challenges. Social scientists have suggested that nanotechnology's social issues should be understood and assessed not simply as "downstream" risks or impacts. Rather, the challenges should be factored into "upstream" research and decision making in order to ensure technology development that meets social objectives

Many social scientists and organizations in civil society suggest that technology assessment and governance should also involve public participation Some observers suggest that nanotechnology will build incrementally, as did the 18-19th century industrial revolution, until it gathers pace to drive a nanotechnological revolution that will radically reshape our economies, our labour markets, international trade, international relations, social structures, civil liberties, our relationship with the natural world and even what we understand to be human. Others suggest that it may be more accurate to describe change driven by nanotechnology as a "technological tsunami". Just like a tsunami, analysts warn that rapid nanotechnology-driven change will necessarily have profound disruptive impacts. As the APEC Centre for Technology Foresight observes: If nanotechnology is going to revolutionize manufacturing, health care, energy supply, communications and probably defence, then it will transform labour and the workplace, the medical system, the transportation and power infrastructures and the military. None of these latter will be changed without significant social disruption.

Those concerned with the negative impact of nanotechnology suggest that it will simply exacerbate problems stemming from existing socio-economic inequity and unequal distributions of power, creating greater inequities between rich and poor through an inevitable nano-divide (the gap between those who control the new nanotechnologies and those whose products, services or labour are displaced by them). Analysts suggest the possibility that nanotechnology has the potential to destabilize international relations through a nano arms race and the increased potential for bioweaponry; thus, providing the tools for ubiquitous surveillance with significant implications for civil liberties. Also, many critics believe it might break down the barriers between life and non-life through nanobiotechnology, redefining even what it means to be human. Nanoethicists posit that such a transformative technology could exacerbate the divisions of rich and poor – the so-called "nano divide." However nanotechnology makes the production of technology, *e.g.* computers, cellular phones, health technology etcetera, cheaper and therefore accessible to the poor.

In fact, many of the most enthusiastic proponents of nanotechnology, such as transhumanists, see the nascent science as a mechanism to changing human nature itself – going beyond curing disease and enhancing human characteristics. Discussions on nanoethics have been hosted by the federal government, especially in the context of "converging technologies" – a catch-phrase used to refer to nano, biotech, information technology, and cognitive science.

Possible Military Applications

Societal risks from the use of nanotechnology have also been raised. On the instrumental level, these include the possibility of military applications of nanotechnology (for instance, as in implants and other means for soldier enhancement like those being developed at the Institute for Soldier Nanotechnologies at MIT) as well as enhanced surveillance capabilities through nano-sensors. There is also the possibility of nanotechnology being used to develop chemical weapons and because they will be able to develop the chemicals from the atom scale up, critics fear that chemical weapons developed from nano particles will be more dangerous than present chemical weapons.

Intellectual Property Issues

On the structural level, critics of nanotechnology point to a new world of ownership and corporate control opened up by nanotechnology. The claim is that, just as biotechnology's ability to manipulate genes went hand in hand with the patenting of life, so too nanotechnology's ability to manipulate molecules has led to the patenting of matter.

The last few years has seen a gold rush to claim patents at the nanoscale. Over 800 nano-related patents were granted in 2003, and the numbers are increasing year to year. Corporations are already taking out broad-ranging patents on nanoscale discoveries and inventions. For example, two corporations, NEC and IBM, hold the basic patents on carbon nanotubes, one of the current cornerstones of nanotechnology.

Carbon nanotubes have a wide range of uses, and look set to become crucial to several industries from electronics and computers, to strengthened materials to drug delivery and diagnostics. Carbon nanotubes are poised to become a major traded commodity with the potential to replace major conventional raw materials. However, as their use expands, anyone seeking to (legally) manufacture or sell carbon nanotubes, no matter what the application, must first buy a license from NEC or IBM.

ENERGY

The most advanced nanotechnology projects related to energy are: storage, conversion, manufacturing improvements by reducing materials and process rates, energy saving (by better thermal insulation for example), and enhanced renewable energy sources.

Reduction of Energy Consumption

A reduction of energy consumption can be reached by better insulation systems, by the use of more efficient lighting or combustion systems, and by use of lighter and stronger materials in the transportation sector. Currently used light bulbs only convert approximately 5 per cent of the electrical energy into light. Nanotechnological approaches like light-emitting diodes (LEDs) or quantum caged atoms (QCAs) could lead to a strong reduction of energy consumption for illumination.

Increasing the Efficiency of Energy Production

Today's best solar cells have layers of several different semiconductors stacked together to absorb light at different energies but they still only manage to use 40 per cent of the Sun's energy.

Commercially available solar cells have much lower efficiencies (15-20 per cent). Nanotechnology could help increase the efficiency of light conversion by using nanostructures with a continuum of bandgaps. The degree of efficiency of the internal combustion engine is about 30-40 per cent at the moment.

Nanotechnology could improve combustion by designing specific catalysts with maximized surface area. In 2005, scientists at the University of Toronto developed a spray-on nanoparticle substance that, when applied to a surface, instantly transforms it into a solar collector.

The Use of More Environmentally Friendly Energy Systems

An example for an environmentally friendly form of energy is the use of fuel cells powered by hydrogen, which is ideally produced by renewable energies. Probably the most prominent nanostructured material in fuel cells is the catalyst consisting of carbon supported noble metal particles with diametres of 1-5 nm. Suitable materials for hydrogen storage contain a large number of small nanosized pores.

Therefore many nanostructured materials like nanotubes, zeolites or alanates are under investigation.

Nanotechnology can contribute to the further reduction of combustion engine pollutants by nanoporous filters, which can clean the exhaust mechanically, by catalytic converters based on nanoscale noble metal particles or by catalytic coatings on cylinder walls and catalytic nanoparticles as additive for fuels.

Recycling of Batteries

Because of the relatively low energy density of batteries the operating time is limited and a replacement or recharging is needed. The huge number of spent batteries and accumulators represent a disposal problem. The use of batteries with higher energy content or the use of rechargeable batteries or supercapacitors with higher rate of recharging using nanomaterials could be helpful for the battery disposal problem.

INFORMATION AND COMMUNICATION

Current high-technology production processes are based on traditional top down strategies, where nanotechnology has already been introduced silently. The critical length scale of integrated circuits is already at the nanoscale (50 nm and below) regarding the gate length of transistors in CPUs or DRAM devices.

Memory Storage

Electronic memory designs in the past have largely relied on the formation of transistors however research into crossbar switch based electronics have offered an alternative using reconfigurable interconnections between vertical and horizontal wiring arrays to create ultra high density memories.

Two leaders in this area are Nantero which has developed a carbon nanotube based crossbar memory called Nano-RAM and Hewlett-Packard which has proposed the use of memristor material as a future replacement of Flash memory.

Novel Semiconductor Devices

An example of such novel devices is based on spintronics.The dependence of the resistance of a material (due to the spin of the electrons) on an external field is called magnetoresistance. This effect can be significantly amplified (GMR - Giant Magneto-Resistance) for nanosized objects, for example when two ferromagnetic layers are separated by a nonmagnetic layer, which is several nanometres thick (*e.g.* Co-Cu-Co).

The GMR effect has led to a strong increase in the data storage density of hard disks and made the gigabyte range possible. The so called tunneling magnetoresistance (TMR) is very similar to GMR and based on the spin dependent tunneling of electrons through adjacent ferromagnetic layers.

Both GMR and TMR effects can be used to create a non-volatile main memory for computers, such as the so called magnetic random access memory or MRAM. In 1999, the ultimate CMOS transistor developed at the Laboratory for Electronics and Information Technology in Grenoble, France, tested the limits of the principles of the MOSFET transistor with a diametre of 18 nm (approximately 70 atoms placed side by side). This was almost one tenth the size of the smallest industrial transistor in 2003. It enabled the theoretical integration of seven billion junctions on a □1 coin. However, the CMOS transistor, which was created in 1999, was not a simple research experiment to study how CMOS technology functions, but rather a demonstration of how this technology functions now that we ourselves are getting ever closer to working on a molecular scale. Today it would be impossible to master the coordinated assembly of a large number of these transistors on a circuit and it would also be impossible to create this on an industrial level.

Novel Optoelectronic Devices

In the modern communication technology traditional analog electrical devices are increasingly replaced by optical or optoelectronic devices due to their enormous bandwidth and capacity, respectively. Two promising examples are photonic crystals and quantum dots. Photonic crystals are materials with a periodic variation in the refractive index with a lattice constant that is half the wavelength of the light used.

They offer a selectable band gap for the propagation of a certain wavelength, thus they resemble a semiconductor, but for light or photons instead of electrons. Quantum dots are nanoscaled objects, which can be used, among many other things, for the construction of lasers. The advantage of a quantum dot laser over the traditional semiconductor laser is that their emitted wavelength depends on the diametre of the dot. Quantum dot lasers are cheaper and offer a higher beam quality than conventional laser diodes.

Displays

The production of displays with low energy consumption could be accomplished using carbon nanotubes (CNT). Carbon nanotubes be electrically conductive and due to their small diametre of several nanometres, they can be used as field emitters with extremely high efficiency for field emission displays (FED). The principle of operation resembles that of the cathode ray tube, but on a much smaller length scale.

Quantum Computers

Entirely new approaches for computing exploit the laws of quantum mechanics for novel quantum computers, which enable the use of fast quantum algorithms. The Quantum computer will have quantum bit memory space termed qubit for several computations at the same time.

HEAVY INDUSTRY

Aerospace

Lighter and stronger materials will be of immense use to aircraft manufacturers, leading to increased performance. Spacecraft will also benefit, where weight is a major factor. Nanotechnology thereby helps to reduce the size of equipments used.

Hang gliders halve their weight while increasing their strength and toughness through the use of nanotech materials. Nanotech is lowering the mass of supercapacitors that will increasingly be used to give power to assistive electrical motors for launching hang gliders off flatland to thermal-chasing altitudes.

Refineries

Using nanotech applications, refineries producing materials such as steel and aluminium will be able to remove any impurities in the materials they create.

Vehicle Manufacturers

Much like aerospace, lighter and stronger materials will be useful for creating vehicles that are both faster and safer. Combustion engines will also benefit from parts that are more hard-wearing and more heat-resistant.

CONSUMER GOODS

Nanotechnology is already impacting the field of consumer goods, providing products with novel functions ranging from easy-to-clean to scratch-resistant. Modern textiles are wrinkle-resistant and stain-repellent; in the mid-term clothes will become "smart", through embedded "wearable electronics". Already in use are different nanoparticle improved products.

Especially in the field of cosmetics, such novel products have a promising potential.

Foods

Nanotechnology can be applied in the production, processing, safety and packaging of food. A nanocomposite coating process could improve food packaging by placing anti-microbial agents directly on the surface of the coated film.

Nanocomposites could increase or decrease gas permeability of different fillers as is needed for different products.

They can also improve the mechanical and heat-resistance properties and lower the oxygen transmission rate. Research is being performed to apply nanotechnology to the detection of chemical and biological substances for sensing biochemical changes in foods.

Household

The most prominent application of nanotechnology in the household is self-cleaning or "easy-to-clean" surfaces on ceramics or glasses. Nanoceramic particles have improved the smoothness and heat resistance of common household equipment such as the flat iron.

Optics

The first sunglasses using protective and antireflective ultrathin polymer coatings are on the market. For optics, nanotechnology also offers scratch resistant surface coatings based on nanocomposites. Nano-optics could allow for an increase in precision of pupil repair and other types of laser eye surgery.

Textiles

The use of engineered nanofibres already makes clothes water- and stain-repellent or wrinkle-free. Textiles with a nanotechnological finish can be washed less frequently and at lower temperatures. Nanotechnology has been used to integrate tiny carbon particles membrane and guarantee full-surface protection from electrostatic charges for the wearer.

Cosmetics

One field of application is in sunscreens. The traditional chemical UV protection approach suffers from its poor long-term stability. A sunscreen based on mineral nanoparticles such as titanium dioxide offer several advantages. Titanium oxide nanoparticles have a comparable UV protection property as the bulk material, but lose the cosmetically undesirable whitening as the particle size is decreased.

APPLICATIONS OF NANOTUBE

The properties of carbon nanotubes have caused researchers and companies to consider using them in several fields. For example, because carbon nanotubes have the highest strength to weight ratio of any known material, researchers at NASA are combining nanotubes with other materials into composites that can be used to build lightweight spacecraft.

Another property of nanotubes is that they can easily penetrate membrances such as cell walls. In fact, nanotubes long, narrow shape make them look like miniature needles, so it makes sense that they can function like a needle at the cellular level. Medical researchers are using this property by attaching molecules that are attracted to cancer cells to nanotubes to deliver drugs directly to the diseased cells.

Another interesting property of nanotubes is that their electrical resistance changes significantly when other molecules attach themselves to the carbon atoms. Companies are using this property to develop sensors that can detect chemical vapours such as carbon monoxide or biological molecules. These are just a few of the potential uses of carbon nanotubes The following survey of carbon nanotube applications introduces these and many other uses.

A Survey of Carbon Nanotube Applications Under Development

Researchers and companies are working to use carbon nanotubes in various fields. The list below introduces many of these uses. Strong, lightweight composites of carbon nanotubes and other materials that can be used to build lightweight spacecraft.

Cables made from carbon nanotubes strong enough to be used for the Space Elevator to drastically reduce the cost of lifting people and materials into orbit. Taking advantage of nanotubes ability to enter cancer cells by attaching targeting molecules which have an affinity to cancer cells as well as anti-cancer drugs to the nanotubes which safety transports an anti-cancer drug through the bloodstream to the tumor.

Stronger bicycle components made by adding carbon nanotubes to a matrix of carbon fibres. Improve the healing process for broken bones by providing a carbon nanotube scaffold for new bone material to grow on. Sensors using carbon nanotube detection elements capable of detecting a range of chemical vapours. These sensors depend upon the fact that the resistance of a carbon nanotube changes in the presence of a chemical vapour.

IMPLICATIONS

Due to the far-ranging claims that have been made about potential applications of nanotechnology, a number of serious concerns have been raised about what effects these will have on our society if realised, and what action if any is appropriate to mitigate these risks.

There are possible dangers that arise with the development of nanotechnology. The Centre for Responsible Nanotechnology suggests that new developments could result, among other things, in untraceable weapons of mass destruction, networked cameras for use by the government, and weapons developments fast enough to destabilize arms races ("Nanotechnology Basics"). One area of concern is the effect that industrial-scale manufacturing and use of nano-materials would have on human health and the environment, as suggested by nanotoxicology research. Groups such as the Centre for Responsible Nanotechnology have advocated that nanotechnology should be specially regulated by governments for these reasons. Others counter that overregulation would stifle scientific research and the development of innovations which could greatly benefit mankind.

Other experts, including director of the Woodrow Wilson Centre's Project on Emerging Nanotechnologies David Rejeski, have testified[28] that successful commercialization depends on adequate oversight, risk research strategy, and public engage-ment. More recently local municipalities have passed (Berkeley, CA) or are considering ordinances requiring nanomaterial manufacturers to disclose the known risks of their products.

The National Institute for Occupational Safety and Health is conducting research on how nanoparticles interact with the body's systems and how workers might be exposed to nano-sized particles in the manufacturing or industrial use of nanomaterials. NIOSH offers interim guidelines for working with nanomaterials consistent with the best scientific knowledge. In "The Consumer Product Safety Commission and Nanotechnology," Felcher suggests that the Consumer Product Safety Commission, which is charged with protecting the public against unreasonable risks of injury or death associated with consumer products, is ill-equipped to oversee the safety of complex, high-tech products made using nanotechnology.

Longer-term concerns centre on the implications that new technologies will have for society at large, and whether these could possibly lead to either a post scarcity economy, or alternatively exacerbate the wealth gap between developed and developing nations.

The effects of nanotechnology on the society as a whole, on human health and the environment, on trade, on security, on food systems and even on the definition of "human", have not been characterized or politicized.

Health and Environmental Concerns

Some of the recently developed *nanoparticle* products may have unintended consequences. Researchers have discovered that silver nanoparticles used in socks to reduce foot odour are being released in the wash with possible negative consequences.[33] Silver nanoparticles, which are bacteriostatic, may then destroy beneficial bacteria which are important for breaking down organic matter in waste treatment plants or farms. A study at the University of Rochester found that when rats breathed in nanoparticles, the particles settled in the brain and lungs, which led to significant increases in biomarkers for inflammation and stress response. A major study published more recently in Nature nanotechnology suggests some forms of carbon nanotubes – a poster child for the "nanotechnology revolution" – could be as harmful as asbestos if inhaled in sufficient quantities.

Anthony Seaton of the Institute of Occupational Medicine in Edinburgh, Scotland, who contributed to the article on carbon nanotubes said "We know that some of them probably have the potential to cause mesothelioma. So those sorts of materials need to be handled very carefully." In the absence of specific nano-regulation forthcoming from governments, Paull and Lyons have called for an exclusion of engineered nanoparticles from organic food.

Regulation

Calls for tighter regulation of nanotechnology have occurred alongside a growing debate related to the human health and safety risks associated with nanotechnology. Further, there is significant debate about who is responsible for the regulation of nanotechnology. While some non-nanotechnology specific regulatory agencies currently cover some products and processes (to varying degrees) – by "bolting on" nanotechnology to existing regulations – there are clear gaps in these regimes.

In “Nanotechnology Oversight: An Agenda for the Next Administration,” former EPA deputy administrator J. Clarence Davies lays out a clear regulatory roadmap for the next presidential administration and describes the immediate and longer term steps necessary to deal with the current shortcomings of nanotechnology oversight. Stakeholders concerned by the lack of a regulatory framework to assess and control risks associated with the release of nanoparticles and nanotubes have drawn parallels with bovine spongiform encephalopathy (‘mad cow’s disease), thalidomide, genetically modified food, nuclear energy, reproductive technologies, biotechnology, and asbestosis.

The Royal Society report identified a risk of nanoparticles or nanotubes being released during disposal, destruction and recycling, and recommended that “manufacturers of products that fall under extended producer responsibility regimes such as end-of-life regulations publish procedures outlining how these materials will be managed to minimize possible human and environmental exposure”. Reflecting the challenges for ensuring responsible life cycle regulation, the Institute for Food and Agricultural Standards has proposed standards for nanotechnology research and development should be integrated across consumer, worker and environmental standards. They also propose that NGOs and other citizen groups play a meaningful role in the development of these standards. The term “manufacturers” includes persons and businesses that produce nanotubes in California, or import carbon nanotubes into California for sale.

The purpose of this information request will be to identify information gaps and to develop information about carbon nanotubes, an important emerging nanomaterial.

SEMICONDUCTOR FABRICATION

Moore’s Law, Optical Lithography and the Search for Alternatives.

Computer chips (and the silicon based transistors within them) are rapidly shrinking according to a predictable formula. According to the Semiconductor Industry Association’s extrapolation of formulas such as this one (SIA roadmap), it is expected that the sizes of circuits within our chips will reach the size of only a few atoms in about 20 years.

Since almost all of our modern computers are made from silicon semiconductor transistors patterned and carved by light (photolithography), the shrinking of circuits predicted by the SIA may not be the most economical method for the future. An enormous amount of money has been invested in the semiconductor industry in order to consistently shrink and improve our semiconductor electronics. Smaller circuits require less energy, operate more quickly and, of course, take up less space.

Thus, Moore's law has been adhered to since computers first became commercially available. However, this simple shrinking of components can not continue for much longer. As transistors such as the Metal-Oxide Semiconductor Field Effect Transistor is made smaller, both its properties and manufacturing expense change with the scale. Currently, Ultraviolet light is used to create the silicon circuits with a lateral resolution around 200 nm (the wavelength of ultraviolet light). As the circuits shrink below 100 nm new fabrication methods must be created, resulting in increasing costs.

Furthermore, once the circuit size reaches only a few nanometres, quantum effects such as tunneling begin to become important, which drastically changes the ability for the computers to function normally. Thus, novel methods for computer chip fabrication have been and are being intensely sought by microchip manufactures.

MOLECULAR NANOTECHNOLOGY: A LONG-TERM VIEW

Molecular nanotechnology, sometimes called molecular manufacturing, describes engineered nanosystems (nanoscale machines) operating on the molecular scale. Molecular nanotechnology is especially associated with the molecular assembler, a machine that can produce a desired structure or device atom-by-atom using the principles of mechanosynthesis. Manufacturing in the context of productive nanosystems is not related to, and should be clearly distinguished from, the conventional technologies used to manufacture nanomaterials such as carbon nanotubes and nanoparticles. When the term "nanotechnology" was independently coined and popularized by Eric Drexler (who at the time was unaware of an earlier usage by Norio Taniguchi) it referred to a future manufacturing technology based on molecular machine systems. The premise was that molecular scale biological analogies of traditional machine components demonstrated molecular machines were possible: by the countless examples found in biology, it is known that sophisticated, stochastically optimised biological machines can be produced.

It is hoped that developments in nanotechnology will make possible their construction by some other means, perhaps using biomimetic principles. However, Drexler and other researchers have proposed that advanced nanotechnology, although perhaps initially implemented by biomimetic means, ultimately could be based on mechanical engineering principles, namely, a manufacturing technology based on the mechanical functionality of these components (such as gears, bearings, motors, and structural members) that would enable programmable, positional assembly to atomic specification. The physics and engineering performance of exemplar designs were analysed in Drexler's book *Nanosystems*. In general it is very difficult to assemble devices on the atomic scale, as all one has to position atoms on other atoms of comparable size and stickiness. Another view, put forth by Carlo Montemagno, is that future nanosystems will be hybrids of silicon technology and biological molecular machines. Yet another view, put forward by the late Richard Smalley, is that mechanosynthesis is impossible due to the difficulties in mechanically manipulating individual molecules. This led to an exchange of letters in the ACS publication Chemical and Engineering News in 2003.

Though biology clearly demonstrates that molecular machine systems are possible, non-biological molecular machines are today only in their infancy. Leaders in research on non-biological molecular machines are Dr. Alex Zettl and his colleagues at Lawrence Berkeley Laboratories and UC Berkeley. They have constructed at least three distinct molecular devices whose motion is controlled from the desktop with changing voltage: a nanotube nanomotor, a molecular actuator, and a nanoelectromechanical relaxation oscillator. An experiment indicating that positional molecular assembly is possible was performed by Ho and Lee at Cornell University in 1999. They used a scanning tunneling microscope to move an individual carbon monoxide molecule (CO) to an individual iron atom (Fe) sitting on a flat silver crystal, and chemically bound the CO to the Fe by applying a voltage.

NANOMATERIALS

The nanomaterials field includes subfields which develop or study materials having unique properties arising from their nanoscale dimensions.

- Interface and colloid science has given rise to many materials which may be useful in nanotechnology, such as carbon nanotubes and other fullerenes, and various nanoparticles and nanorods. Nanomaterials with fast ion transport are related also to nanoionics and nanoelectronics.
- Nanoscale materials can also be used for bulk applications; most present commercial applications of nanotechnology are of this flavour.
- Progress has been made in using these materials for medical applications.
- Nanoscale materials are sometimes used in solar cells which combats the cost of traditional Silicon solar cells
- Development of applications incorporating semiconductor nanoparticles to be used in the next generation of products, such as display technology, lighting, solar cells and biological imaging.

BOTTOM-UP APPROACHES

These seek to arrange smaller components into more complex assemblies.

- DNA nanotechnology utilizes the specificity of Watson–Crick basepairing to construct well-defined structures out of DNA and other nucleic acids.
- Approaches from the field of "classical" chemical synthesis also aim at designing molecules with well-defined shape (*e.g.* bis-peptides).

- More generally, molecular self-assembly seeks to use concepts of supramolecular chemistry, and molecular recognition in particular, to cause single-molecule components to automatically arrange themselves into some useful conformation.
- Atomic force microscope tips can be used as a nanoscale "write head" to deposit a chemical upon a surface in a desired pattern in a process called dip pen nanolithography. This technique fits into the larger subfield of nanolithography.

TOP-DOWN APPROACHES

These seek to create smaller devices by using larger ones to direct their assembly.

- Many technologies that descended from conventional solid-state silicon methods for fabricating microprocessors are now capable of creating features smaller than 100 nm, falling under the definition of nanotechnology. Giant magnetoresistance-based hard drives already on the market fit this description, as do atomic layer deposition (ALD) techniques. Peter Grünberg and Albert Fert received the Nobel Prize in Physics in 2007 for their discovery of Giant magnetoresistance and contributions to the field of spintronics.
- Solid-state techniques can also be used to create devices known as nanoelectromechanical systems or NEMS, which are related to microelectromechanical systems or MEMS.
- Focused ion beams can directly remove material, or even deposit material when suitable pre-cursor gasses are applied at the same time. For example, this technique is used routinely to create sub-100 nm parts of material for analysis in Transmission electron microscopy.
- Atomic force microscope tips can be used as a nanoscale "write head" to deposit a resist, which is then followed by an etching process to remove material in a top-down method.

FUNCTIONAL APPROACHES

These seek to develop components of a desired functionality without regard to how they might be assembled.

- Molecular scale electronics seeks to develop molecules with useful electronic properties. These could then be used as single-molecule components in a nanoelectronic device.
- Synthetic chemical methods can also be used to create synthetic molecular motors, such as in a so-called nanocar.

BIOMIMETIC APPROACHES

- Bionics or biomimicry seeks to apply biological methods and systems found in nature, to the study and design of engineering systems and modern technology. Biomineralization is one example of the systems studied.

- Bionanotechnology is the use of biomolecules for applications in nanotechnology, including use of viruses.

SPECULATIVE

These subfields seek to anticipate what inventions nanotechnology might yield, or attempt to propose an agenda along which enquiry might progress. These often take a big-picture view of nanotechnology, with more emphasis on its societal implications than the details of how such inventions could actually be created.

- Molecular nanotechnology is a proposed approach which involves manipulating single molecules in finely controlled, deterministic ways. This is more theoretical than the other subfields and is beyond current capabilities.
- Nanorobotics centers on self-sufficient machines of some functionality operating at the nanoscale. There are hopes for applying nanorobots in medicine, but it may not be easy to do such a thing because of several drawbacks of such devices. Nevertheless, progress on innovative materials and methodologies has been demonstrated with some patents granted about new nanomanufacturing devices for future commercial applications, which also progressively helps in the development towards nanorobots with the use of embedded nanobioelectronics concepts.
- Productive nanosystems are "systems of nanosystems" which will be complex nanosystems that produce atomically precise parts for other nanosystems, not necessarily using novel nanoscale-emergent properties, but well-understood fundamentals of manufacturing. Because of the discrete (*i.e.* atomic) nature of matter and the possibility of exponential growth, this stage is seen as the basis of another industrial revolution. Mihail Roco, one of the architects of the USA's National Nanotechnology Initiative, has proposed four states of nanotechnology that seem to parallel the technical progress of the Industrial Revolution, progressing from passive nanostructures to active nanodevices to complex nanomachines and ultimately to productive nanosystems.
- Programmable matter seeks to design materials whose properties can be easily, reversibly and externally controlled though a fusion of information science and materials science.
- Due to the popularity and media exposure of the term nanotechnology, the words picotechnology and femtotechnology have been coined in analogy to it, although these are only used rarely and informally.

4

Environmental Health and Safety Impact of Nanotechnology

HEALTH IMPACT OF NANOTECHNOLOGY

The health impact of nanotechnology are the possible effects that the use of nanotechnological materials and devices will have on human health. As nanotechnology is an emerging field, there is great debate regarding to what extent nanotechnology will benefit or pose risks for human health. Nanotechnology's health impact can be split into two aspects: the potential for nanotechnological innovations to have medical applications to cure disease, and the potential health hazards posed by exposure to nanomaterials.

NANOTOXICOLOGY

The extremely small size of nanomaterials also means that they are much more readily taken up by the human body than larger sized particles. How these nanoparticles behave inside the body is one of the issues that needs to be resolved. The behaviour of nanoparticles is a function of their size, shape and surface reactivity with the surrounding tissue. They could cause overload on phagocytes, cells that ingest and destroy foreign matter, thereby triggering stress reactions that lead to inflammation and weaken the body's defence against other pathogens. Apart from what happens if non-degradable or slowly degradable nanoparticles accumulate in organs, another concern is their potential interaction with biological processes inside the body: because of their large surface, nanoparticles on exposure to tissue and fluids will immediately adsorb onto their surface some of the macromolecules they encounter. This may, for instance, affect the regulatory mechanisms of enzymes and other proteins.

Other properties of nanomaterials that influence toxicity include: chemical composition, shape, surface structure, surface charge, aggregation and solubility, and the presence or absence of functional groups of other chemicals. The large number of variables influencing toxicity means that it is difficult to generalise about health risks associated with exposure to nanomaterials – each new nanomaterial must be assessed individually and all material properties must be taken into account.

California

In October 2008, the Department of Toxic Substances Control (DTSC), within the California Environmental Protection Agency, announced its intent to request information regarding analytical test methods, fate and transport in the environment, and other relevant information from manufacturers of carbon nanotubes. These sections were added as a result of the adoption of Assembly Bill AB 289 (2006). They are intended to make information on the fate and transport, detection and analysis, and other information on chemicals more available. The law places the responsibility to provide this information to the Department on those who manufacture or import the chemicals.

On January 22, 2009, a formal information request letter was sent to manufacturers who produce or import carbon nanotubes in California, or who may export carbon nanotubes into the State.

This letter constitutes the first formal implementation of the authorities placed into statute by AB 289 and is directed to manufacturers of carbon nanotubes, both industry and academia within the State, and to manufacturers outside California who export carbon nanotubes to California. This request for information must be met by the manufacturers within one year. DTSC is waiting for the upcoming January 22, 2010 deadline for responses to the data call-in. The California Nano Industry Network and DTSC hosted a full-day symposium on November 16, 2009 in Sacramento, CA. This symposium provided an opportunity to hear from nanotechnology industry experts and discuss future regulatory considerations in California. DTSC is expanding the Specific Chemical Information Call-in to members of the nanometal oxides.

Nanomedicine

Nanomedicine is the medical application of nanotechnology. The approaches to nanomedicine range from the medical use of nanomaterials, to nanoelectronic biosensors, and even possible future applications of molecular nanotechnology. Current problems for nanomedicine involve understanding the issues related to toxicity and environmental impact of nanoscale materials.

Nanomedicine research is directly funded, with the US National Institutes of Health in 2005 funding a five-year plan to set up four nanomedicine centres. In April 2006, the journal Nature Materials estimated that 130 nanotech-based drugs and delivery systems were being developed worldwide.

Nanomedicine seeks to deliver a set of research tools and clinical devices in the near future. The National Nanotechnology Initiative expects new commercial applications in the pharmaceutical industry that may include advanced drug delivery systems, new therapies, and *in vivo* imaging. Neuro-electronic interfaces and other nanoelectronics-based sensors are another active goal of research. Further down the line, the speculative field of molecular nanotechnology believes that cell repair machines could revolutionize medicine and the medical field. Nanomedicine is a large industry, with nanomedicine sales reaching $6.8 billion in 2004. With over 200 companies and 38 products worldwide, a minimum of $3.8 billion in nanotechnology R&D is being invested every year. As the nanomedicine industry continues to grow, it is expected to have a significant impact on the economy.

Curing Diseases

Currently, nanotech gene therapy has been able to kill ovarian cancer in mice while avoiding the side effects of *cisplatin* and *paclitaxel*; it is speculated that this technology could save 15000 women in the United States each year if the treatment proves effective and safe in humans.

Research on nanoelectronics-based cancer diagnostics could lead to tests that can be done in pharmacies. The results promise to be highly accurate and the product promises to be inexpensive. They could take a very small amount of blood and detect cancer anywhere in the body in about five minutes, with a sensitivity that is a thousand times better than in a conventional laboratory test. These devices that are built with nanowires to detect cancer proteins; each nanowire detector is primed to be sensitive to a different cancer marker. The biggest advantage of the nanowire detectors is that they could test for anywhere from ten to one hundred similar medical conditions without adding cost to the testing device.

HUMAN, ENVIRONMENTAL HEALTH AND SAFETY

Studies of the health impact of airborne particles generally shown that for toxic materials, smaller particles are more toxic. This is due in part to the fact that, given the same mass per volume, the dose in terms of particle numbers increases as particle size decreases. Based upon available data, it has been argued that current risk assessment methodologies are not suited to the hazards associated with nanoparticles; in particular, existing toxicological and eco-toxicological methods are not up to the task; exposure evaluation needs to be expressed as quantity of nanoparticles and/or surface area rather than simply mass; equipment for routine detecting and measuring nanoparticles in air, water, or soil is inadequate; and very little is known about the physiological responses to nanoparticles.

Regulatory bodies in the U.S. as well as in the EU have concluded that nanoparticles form the potential for an entirely new risk and that it is necessary to carry out an extensive analysis of the risk. The challenge for regulators is whether a matrix can be developed which would identify nanoparticles and more complex nanoformulations which are likely to have special toxicological properties or whether it is more reasonable for each particle or formulation to be tested separately. The International Council on Nanotechnology maintains a database and Virtual Journal of scientific papers on environmental, health and safety research on nanoparticles. The database currently has over 2000 entries indexed by particle type, exposure pathway and other criteria.

The *Project on Emerging Nanotechnologies* (PEN) currently lists 807 products that manufacturers have voluntarily identified that use nanotechnology. No labeling is required by the FDA so that number could be significantly higher. "The use of nanotechnology in consumer products and industrial applications is growing rapidly, with the products listed in the PEN inventory showing just the tip of the iceberg" according to PEN Project Director David Rejeski. A list of those products that have been voluntarily disclosed by their manufacturers is located here. The Material Safety Data Sheet that must be issued for certain materials often does not differentiate between bulk and nanoscale size of the material in question and even when it does these MSDS are advisory only.

DEMOCRATIC GOVERNANCE

Many argue that government has a responsibility to provide opportunities for the public to be involved in the development of new forms of science and technology. Community engagement can be achieved through various means or mechanisms. An online journal article identifies traditional approaches such as referenda, consultation documents, and advisory committees that include community members and other stakeholders. Other conventional approaches include public meetings and "closed" dialog with stakeholders.

More contemporary engagement processes that have been employed to include community members in decisions about nanotechnology include citizens' juries and consensus conferences. Leach and Scoones argue that since that "most debates about science and technology options involve uncertainty, and often ignorance, public debate about regulatory regimes is essential." It has been argued that limited nanotechnology labeling and regulation may exacerbate potential human and environmental health and safety issues associated with nanotechnology, and that the development of comprehensive regulation of nanotechnology will be vital to ensure that the potential risks associated with the research and commercial application of nanotechnology do not overshadow its potential benefits.

Regulation may also be required to meet community expectations about responsible development of nanotechnology, as well as ensuring that public interests are included in shaping the development of nanotech-nology. Community education, engagement and consultation tend to occur "downstream": once there is at least a moderate level of awareness, and often during the process of disseminating and adapting technologies.

"Upstream" engagement, by contrast, occurs much earlier in the innovation cycle and involves: "dialogue and debate about future technology options and pathways, bringing the often expert-led approaches to horizon scanning, technology foresight and scenario planning to involve a wider range of perspectives and inputs." Daniel Sarewitz Director of Arizona State University's Consortium on Science, Policy and Outcomes, argues that "by the time new devices reach the stage of commercialization and regulation, it is usually too late to alter them to correct problems." The stance that the research, development and use of nanotechnology should be subject to control by the public sector is sometimes referred to as nanosocialism.

NEWNESS

The question of whether nanotechnology represents something 'new' must be answered to decide how best nanotechnology should be regulated. The Royal Society recommended that the UK government assess chemicals in the form of nanoparticles or nanotubes as new substances. Subsequent to this, in 2007 a coalition of over forty groups called for nanomaterials to be classified as new substances, and regulated as such. Despite these recommendations, chemicals comprising nanoparticles that have previously been subject to assessment and regulation may be exempt from regulation, regardless of the potential for different risks and impacts. In contrast, nanomaterials are often recognized as 'new' from the perspective of *intellectual property rights* (IPRs), and as such are commercially protected via patenting laws.

NEW REGULATORY FRAMEWORK, OR ADAPT EXISTING ARRANGEMENTS

There is significant debate about who is responsible for the regulation of nanotechnology. While some non-nanotechnology specific regulatory agencies currently cover some products and processes–by "bolting on" nanotechnology to existing regulations–there are clear gaps in these regimes. This enables some nanotechnology applications to figuratively "slip through the cracks" without being covered by any regulations. An example of this has occurred in the US, and involves nanoparticles of titanium dioxide for use in sunscreen where they create a clearer cosmetic appearance. In this case, the US *Food and Drug Administration* (FDA) reviewed the immediate health effects of exposure to nanoparticles of *titanium dioxide* (TI02) for consumers. However, they did not review its impacts for aquatic ecosystems when the sunscreen rubs off, nor did the EPA, or any other agency. Similarly the Australian equivalent of the FDA, the *Therapeutic Goods Administration* (TGA) approved the use of nanoparticles in sunscreens after a thorough review of the literature, on the basis that although nanoparticles of TI02 and *zinc oxide* (ZN0) in sunscreens do produce free radicals and oxidative DNA damage *in vitro*, such particles were unlikely to pass the dead outer cells of the stratum corneum of human skin; a finding which some academics have argued seemed not to apply the precautionary principle in relation to prolonged use on children with cut skin, the elderly with thin skin, people with diseased skin or use over flexural creases.

Doubts over the TGA's decision were raised with publication of a paper showing that the uncoated anatase form of TI02 used in some Australian sunscreens caused a photocatalytic reaction that degraded the surface of newly installed prepainted steel roofs in places where they came in contact with the sunscreen coated hands of workmen. Such gaps in regulation are likely to continue alongside the development and commercialization of increasingly complex second and third generation nanotech-nologies. Nanomedicines are just beginning to enter drug regulatory processes, but within a few decades could comprise a dominant group within the class of innovative pharmaceuticals, the current thinking of government safety and cost-effectiveness regulators appearing to be that these products give rise to few if any nano-specific issues.

Some academics have challenged that proposition and suggest that nanomedicines may create unique or heightened policy challenges for government systems of cost-effectiveness as well as safety regulation. There are also significant public good aspects to the regulation of nanotechnology, particularly with regard to ensuring that industry involvement in standard-setting does not become a means of reducing competition and that nanotechnology policy and regulation encourages new models of safe drug discovery and development more systematically targeted at the global burden of disease.

The Self-policing wiki notes that self-regulation attempts may well fail, due to the inherent conflict of interest in asking any organization to police itself. If the public becomes aware of this failure, an external, independent organization is often given the duty of policing them, sometimes with highly punitive measures taken against the organization. The Food and Drug Administration notes that it only regulates on the basis of voluntary claims made by the product manufacturer. If no claims are made by a manufacturer, then the FDA may be unaware of nanotechnology being employed.

Yet regulations worldwide still fail to distinguish between materials in their nanoscale and bulk form. This means that nanomaterials remain effectively unregulated; there is no regulatory requirement for nanomaterials to face new health and safety testing or environmental impact assessment prior to their use in commercial products, if these materials have already been approved in bulk form. The health risks of nanomaterials are of particular concern for workers who may face occupational exposure to nanomaterials at higher levels, and on a more routine basis, than the general public.

INTERNATIONAL LAW

There is no international regulation of nanoproducts or the underlying nanotechnology. Nor are there any internationally agreed definitions or terminology for nanotechnology, no internationally agreed protocols for toxicity testing of nanoparticles, and no standardized protocols for evaluating the environmental impacts of nanoparticles. Since products that are produced using nanotechnologies will likely enter international trade, it is argued that it will be necessary to harmonize nanotechnology standards across national borders. There is concern that some countries, most notably developing countries, will be excluded from international standards negotiations. The Institute for Food and Agricultural Standards notes that "developing countries should have a say in international nanotechnology standards development, even if they lack capacity to enforce the standards". Concerns about monopolies and concentrated control and ownership of new nanotechnologies were raised in community workshops in Australia in 2004.

HUMAN-ENVIRONMENT INTERACTION

Human beings are endowed by nature to be reflective and active. Their biological evolution gives them capacity to forge tools and establish an adaptive relationship with nature. In the beginning, human life was more biological than cultural and was somewhat similar to other animals where environmental considerations dictated the place of human residence. In the process of adaptive relationship man gradually evolved tools with the help of which the resources of the environment could be put to use.

The tool making ability developed over a very long period of time as it began with the use of materials locally obtainable. The tools shaped human life such that we witness the emergence and growth of 'cultures'. The different stages of human culture have been identified on the basis of the tools used by them. The earliest was the paleolithic age representing the beginning of the tool industry. In this age humans lived by gathering plant foods and hunting animals.

It was inherent in the nature of the economy of the period that humans could not lead sedentary life and were forced to migrate to new places in search of plant foods and game. This kind of life-style restricted the size of the peregrinating human groups. It can be safely argued that during this phase of human history the environment dictated terms and humans had just started making an effort to modify their dependence on nature. Nonetheless, it is necessary to point out that mobility had led to greater interaction between numerous groups of humans spread over different parts of the world.

It will be not out of place here to delineate the adaptive strategy of the early humans so as to explain his interaction with the environment. For this purpose we focus on southwest France. During the upper Palaeolithic phase

(35000-12000 years ago), the climate of this region was strongly oceanic, with cool summers and mild winters (by Ice age standards) affecting the environment. Summer temperature may have been in the 53.6 °F to 59 °F range, with winter readings around 32 °F. The vegetation-growing season was longer on the open plains to the north and east, and snow cover had retreated considerably.

Thus food resources for large herbivores were now more readily available, perhaps resulting in a much higher density of game animals as well as more plentiful edible foods. This region was marked as a region of diverse food resources. The people were mainly subsisted off Reindeer, but they took wild ox, red deer, bison, ibex, chamois, woolly rhinoceros and mammoth too. Many of these resources were relatively predictable. The large-scale salmon fishing during seasonal runs was a major factor in the evolution of complex hunter-gatherer societies in the region.

Effective exploitation of salmon runs requires not only efficient fishing technology but the services of considerable numbers of people to dry and store the thousands of fresh fish before they spoil. These people extensively used fishhooks and harpoons. The people tended to choose many of their settlement sites with reference to plentiful water supply and good views of the surrounding landscape, so they could observe game. When the people occupied a rock shelter or cave, it invariably faced south, so they could benefit from the sun's rays on cool days.

Some of the largest cave and rock shelter sites lay close to river fords, places, perhaps, where migrating reindeer would cross each year. The relationship between nature and man was redefined with the advent of agriculture. Till the beginning of agriculture, the sources of food had mostly been naturally available products and man had no control over their availability. An important contribution of agriculture has been the cultivation of cereals. The fact is that the shelf-life of cereals is unlimited whereas fruits and meat had very limited shelf-life.

It has been a very significant factor as this property of cereals encouraged accumulation, which perhaps was one of the causes for the introduction and intensification of social stratification. In the beginning agriculture was a highly unreliable source of food, and transition from hunter-gatherer to peasant was not very smooth and was a long drawn process. The development of technology/tools to increase agricultural production was a continuing process in which development of irrigation technology too played an important role. Slowly but surely agriculture became the major source of subsistence and increased productivity contributed towards increase in population.

Initially agriculture was confined to highly favourable locations with natural irrigation. With the growth in population, however, man was forced to migrate to less-favourable locations, necessitating irrigation. The development of irrigation facilities required larger social participation and better management resulting in a transition towards complex society. Furthermore better management of agriculture insured food security and provided humans with surplus time since agriculture was a seasonal activity.

Likewise demand for improved tools and technology for better irrigation to ensure larger production led to depletion of locally available raw materials for tools (for example stone, as man moved away from foothills to open plains). This compelled man to look for other kinds of materials and other locations to augment the supply of raw material for tool making. Meanwhile, the introduction of the wheel had revolutionized movement and encouraged the emergence of wheel-based pottery, a highly specialized occupation.

The gradual development in technology attained another stage as metallurgy developed. The discovery of metallic ores once again redefined the man-environment interaction. The major advantage of metal tools over stone was its reusable character: stone tools once broken could not be used again whereas metal tools could be remoulded. However, the relative scarcity of mineral ores together with the limited capabilities of processing, beginning from procurement to transportation and finally extraction made metal procurement a labour intensive and expensive proposition.

The most important feature of metallurgy was the highly specialized knowledge required and expertise, which made it a full-time occupation. The emergence of such professionals could be sustained only with the availability of agricultural surplus. This led to the emergence of a section of the population not directly involved with the food production. The parasitic character of this section of population gradually liberated from direct dependence on nature and heralded a new era where certain sections of the inhabitants survived solely on their professional knowledge.

The character of agriculture based societies has been defined in terms of complex social stratification with specialization of craft. The growing ability of humans to make use of a variety of environmental resources opened up the possibilities of the exploitation of natural resources for self-benefit. The larger equity based and open community now witnessed a transition towards a rudimentary system of sociopolitico- economic hierarchy. Still, we cannot say that humans were controlling the environment rather the nature of dependence on environment had changed drastically.

The most defined form of control over nature became visible only in the Industrial Age. Unprecedented growth of technology during the Industrial Age (second half of the 18th century to the beginning of 20th century) liberated man from physical labour and an alienation with the natural world gradually set in. The Industrial Age introduced the exploitation of abiotic source of energy (which are not biologically procurable) and gradually replaced human and animal energy as the dominating forms. Since the ancient past thermal energy had been used in direct application, but during the Industrial Age it was used to mechanize tools. The Industrial Age witnessed the conversion of thermal energy to mechanical energy and thus enhanced the possibilities of greater exploitation of natural resources.

The conversion of thermal energy to other forms of energy tremendously increased the overall demand for energy and resulted in a gradual depletion of the sources of energy. Consequently search for newer sources of thermal energy began: hydrocarbons, *i.e.*, coal, petroleum products, etc., were explored and the magnitude of their exploitation widened. Unlike the earlier renewable source of energy like human and animal labour and wood, newer sources of energy *i.e.* hydrocarbons are non-renewable in character or have economically unviable extralong cycles of renewal.

The introduction of non-renewable source of energy redefined the relationship between the environment and humans. In the modern age ever-growing demand for energy coupled with the steady depletion of sources of energy forced man to reconsider priorities and we see the beginning of the movement for 'conservation.' Better technology ensured greater agricultural production which contributed to a rise in life-expectancy and decline in the mortality rate. The resultant increase in the population in real terms was unprecedented.

It is not that human civilization had never witnessed the growth of population in the past, but the magnitude has been very high in the modern age- the nineteenth century. Ferdinand Braudel has attempted to define it in terms of the ecological watershed, *i.e.*, the end of the 'Biological ancien regime'. He writes: 'What was shattered ... with the eighteenth century was a "Biological ancien regime", a set of restrictions, obstacles, structures, proportions and numerical relationships that had hitherto been the norm.

The chief constituents are:

- Number of death roughly equivalent to the number of births
- Very high infant mortality
- Famine
- Chronic undernourishment
- Formidable epidemics

It is rather broader definition to explain the ecological watershed as it traces the causes in a very long-term perspective beginning with the middle ages and at-least the geographical explorations. At this juncture it is necessary to point out that since the ancient past in Europe we could witness the prevalence of anthropocentric social attitudes. The clearest manifestation was seen in the concept of cosmology in ancient discourses. The earth, the abode of humans, was considered at the centre of the universe and was enveloped by seven strata.

All the seven strata were supposed to have emanated from the earth. The growth of capitalism and the breakdown of the 'biological regime' led to an exponential growth in population. Another corollary of excessive exploitation of environmental resources during the Industrial Age has been the growth of democratic values and institutions. In the same era, scientific knowledge along with technological development provided a world vision where technology was portrayed as a solution to all human problems, especially the problem of hunger and poverty.

Moreover, the growth of scientific and technological knowledge furthered the traditional anthropocentric view and the exploitation of the environment gained a fresh momentum that continues unabated till today. The greater use of energy led to major problem of environmental pollution. The greater consumption and generation of energy induced a 'green house effect'.

However, what has been a more bothersome fall-out of this process is the development of materials not naturally available in the world, *i.e.*, polymers. The chemical revolution of the 1930's and 1940's developed an artificial material which was not biodegradable and was thus difficult to destroy and decompose. At the same time, the wider applications of the material in industrial and domestic use and low cost of production encouraged its wider circulation.

However, the problem of decomposition of the material made it a major cause of concern for the scientific community. Similarly, the question of the viability of nuclear fuel as a source of energy has been a major issue of concern. The production of non-natural radioactive substance for energy production has been a major scientific and technological development, but again the decay or the proper and cost effective decomposition of the residue has been a major technological failure.

INCENTIVES FOR ENVIRONMENTALLY BENEFICIAL TECHNOLOGY

An important aspect of managing NT is to encourage its application to environmentally beneficial uses. Most applications of NT are environmentally beneficial in that they reduce the amount of material necessary for a particular purpose. However, there are more specific environmental benefits that NT may make possible. For example, NT materials already have been used to remove toxic materials from soil at contaminated sites. Government initiatives to encourage environmentally beneficial NT also should be used to encourage applications that benefit public health. In fact, failure to treat NT public health and environmental applications equally would be morally and politically untenable. This part discusses four mechanisms that can be used to encourage applying NT to environmental protection and public health: research, tax breaks, acquisition programmes and regulatory advantages. While each of these is worth considering, each has significant problems.

RESEARCH

The most direct way for government to encourage beneficial NT is for the government to conduct the research itself. Alternatively, the government can pay for research, but let others do the actual work under contract. A third possibility—one in which the government has less control over the research—is for the government to award research grants. In 2000, President Clinton started the National Nanotechnology Initiative (NNI), which included $422 million of funds from various agencies. The 21st Century Nanotechnology Research and Development Act, enacted in 2003, provided a statutory base for the NNI. The NNI is an interagency effort under the general direction of the National Science and Technology Council. The NNI included $989 million for research contained in the budgets of 11 agencies in FY 2004.

In 2005, the amount is estimated to increase to $1.08 billion. The President's budget requests $1.05 billion in FY 2006. Of the 2006 request, $230 million is in the Defence Department budget. EPA is requesting $5 million. Overall, 4 per cent of the federal funds go to research on the health and environmental effects of NT and another 4 per cent goes to "education-related activities and research on the broad implications of nanotechnology for society". Several U.S. states have appropriated money for NT research as part of the competition for economic growth. In 2000, California allocated $95 million towards a Nano-Systems Institute at the University of California Los Angeles and Santa Barbara. In 2003, Oregon appropriated $21 million to launch a Nano/Micro Institute, and New York contributed $50 million to a Nanotech Center at SUNY Albany. Many other national governments are funding NT research.

The European Union is spending significant amounts. In 2004, scientists in China published more articles on nanoscience and nanotechnology than scientists in the United States, and China now ranks third behind the United States and Japan in NT patents. A recent survey found the United States, Japan, Germany and South Korea to be the dominant nations in NT research and development. The effectiveness of this research for encouraging environmentally beneficial NT is unknowable now and may never be known. Dramatic successes and failures of government R&D programmes are widely publicized, but overall evaluations of such programmes are rare. In the United States, it is hard to know what portion of government R&D funds go to NT technologies that are directly or indirectly beneficial to health and the environment. This should be a continuing concern of the NNI.

TAX BREAKS

Tax benefits have been an important tool for protecting the environment. The Byzantine U.S. tax code contains a variety of incentives for environmentally beneficial behaviour such as energy conservation. Tax penalties to discourage pollution or other socially undesirable behaviour have been used in the United States, but are more common in European countries. For example, gas taxes are much higher in most OECD countries than in the United States.

Although one can argue that any imaginable provision can find its way into the U.S. tax code, there are at least two significant obstacles to using taxes to encourage environmentally beneficial NT. One obstacle is that there has never been strong support for such use of the tax system. In particular, environmental groups have been cool to economic incentives of any kind. The tax-writing committees in Congress, especially the powerful House Ways and Means Committee, most frequently have taken the position that taxes should be used only to raise general funds for running the government and have opposed using taxes for social purposes.

The Treasury Department usually has supported the committees in this opposition. The fact that some large corporations are engaged in NT research and development does provide a potential base of support for tax subsidies. The other obstacle lies in defining which expenditures would be eligible for tax subsidies or tax penalties. The definition of which products or processes are NT is not altogether clear, and defining which NT efforts are environmentally desirable is much more problematic. The need to make economically important decisions about poorly delineated categories leads to confusion at best, and to corruption at worst.

ACQUISITION PROGRAMMES

The federal and state governments are major consumers of a broad variety of goods and services. This characteristic has occasionally been used to promote social goals. For example, the Clean Air Act says that the General Services Administration should underwrite the incremental cost of purchasing clean fuel vehicles for use in federal agency fleets. In theory, the government could be required to purchase environmentally beneficial NT products. Such a mandate would raise the same definitional problems as tax breaks.

More fundamentally, it would raise a dilemma that has characterized most acquisition proposals, one of cost versus social goal. Government agencies generally behave like rational consumers, searching for the lowest price for a given product. If the products mandated to be purchased are less expensive than comparable products, agencies will buy them in any case and the legislation is unnecessary. If the mandated products are more expensive, the agencies will be reluctant to spend their budget money to fulfill a purpose that is not part of their mission, and the acquisition programme gets quietly undermined by the resistance of purchasing agents in individual agencies. To the extent that government purchasing is done centrally by the GSA, the chances of acquisition programmes succeeding are slightly better, but the basic dynamic remains.

REGULATORY INCENTIVES

There have been a few instances of using regulatory incentives to encourage particular types of technologies. For example, FDA accelerates the review and approval of certain types of drugs that address unmet medical needs. It would be possible for FDA to give some advantage to NT drugs and medical devices if this were considered desirable. It also would be possible to do this under new legislation dealing with NT. TSCA does not lend itself to such incentives because there is a set time limit for EPA to make a decision about a chemical. At the end of the time period, the chemical can be manufactured unless the agency has taken steps to ban or limit its manufacture. How much of an incentive a regulatory provision is depends on how much capital has been invested in the product before the regulatory decision, and on how much time is saved by the regulatory advantage. In the case of something like a new drug, where the development expense typically is large and the regulatory delay could be lengthy, a regulatory leg up can be an important incentive.

NEW INSTITUTIONAL CAPABILITIES

Dealing successfully with NT's adverse effects will require the establishment of new institutions or institutional mechanisms. While necessary to deal with NT, these institutions and mechanisms also are necessary for dealing with a wide variety of other problems and policies. Although they are discussed in the context of NT, their broader relevance should be kept in mind. The four most important functions that require institutional strengthening and innovation are international harmonization, foresight capability, research and public participation.

INTERNATIONAL HARMONIZATION

Any effort to address NT must be international in scope. Every aspect of the new technology is international. Research on NT is being conducted in a large number of countries. Research and development are not limited to OECD nations—South Korea, China, India and Brazil, among others, are major players. Furthermore, most of the firms marketing or likely to market NT products engage in international outsourcing, licensing of firms in other countries, and other actions across national boundaries that make the notion of "country of origin" seem quite outdated. A regulatory regime for NT should have international coordination built into it. The failure of the U.S. TSCA to do this is one of its many weaknesses.

The part of this chapter that discussed creation of a new law described some ways this might be done. Additional ideas are needed. An international coordinating body would greatly strengthen national efforts to manage NT. One possibility is for OECD to establish an International Council on the Effects of Nanotechnology (ICEN). Membership would not be limited to OECD countries and should include UNEP (United Nations Environment Programme) and WHO (World Health Organization). The council would have three major functions:

- Sharing and disseminating information on the health and environmental effects of NT;
- Making non-binding recommendations for government actions to manage NT's effects; and

- Evaluating the actions taken by member states to address the effects of NT.

The last function would be analogous to OECD's evaluations of national environmental programmes. The current international mechanisms for dealing with environmental and technological problems are a crazy-quilt of small, mostly inadequate, institutions and staffs. An international coordinating body for NT could be just an interim step. At some point, the need for a single strong international body to manage all of the environmental treaties, agreements and understandings will become imperative. The need to deal internationally with NT will add one more reason for nations to create such a body.

FORESIGHT CAPABILITY

The U.S. EPA, the National Research Council and others have called for the development of an improved technology forecasting ability to identify potential environmental impacts of emerging technologies, to interact with business in the early stages of technology development to help design out negative impacts, and to support environmentally positive applications. There has been a long, on-going debate about the ability to do forecasting. If the goal is accurate prediction of the future, then most forecasting is doomed to failure.

However, thinking about the future allows one to consider likely options, identify important determinants of the future of a technology, and get a jump on creating relevant policies. The rapid pace of NT makes improved foresight capability a necessity. Several kinds of initiatives are needed. First, EPA and other agencies should establish offices explicitly charged with forecasting—and with encouraging their agency to use forecast results. Second, NSF, EPA and others should fund academic centers dedicated to research that will improve forecasting ability.

A lot of research on how to improve forecasting is needed but forecasting is inherently interdisciplinary, and for this and other reasons, it has been almost entirely neglected by academics. The availability of funding for research—and the existence of institutions that would use the results—would help remedy this neglect. As the NRC has said, many reasons exist to motivate forecasting activities.

Among them is the desire to increase the lead time for making decisions in order to allow more careful analysis of various options, and to increase the chance for broad public participation in decision making. In managing NT's effects, these reasons are particularly relevant. Congress also needs to have the ability to foresee and evaluate new technological developments. In 1995, the Republican Congress eliminated the Congressional Office of Technology Assessment, the only effective institution it had for fulfilling these tasks. It eliminated OTA as a budget-cutting measure, although the total OTA budget was a modest $22 million. Rebuilding Congress' foresight capability is essential if the legislature is to operate effectively in the modern world.

RESEARCH

The role of research in encouraging environmentally beneficial NT. However, there is also a need for research focusing on adverse effects of NT, monitoring instruments, and control methods to prevent adverse effects. Much of the research on effects is either done or paid for by the government. NT presents a variety of unique challenges. It is not clear whether there are good ways to detect NT materials in the ambient environment, or, if they cannot be detected, whether there are ways environmental controls can be imposed. It is not clear which, if any, of the usual control methods will be effective in dealing with NT materials. Following these avenues often leads back to the primary importance of prevention.

The only way to deal with potential adverse effects of NT in most cases will be to design the product or tailor the use of the material so that the NT material does not get into the environment or the human body in the first place. Rejeski has suggested that EPA needs an advanced research capability similar to the military's Defence Advanced Research Projects Agency to develop breakthrough solutions for the "really hard problems". EPA's track record with research of this kind is not good, but creation of a new office with incentives to attract outstanding researchers could change this. It will remain a continuing challenge to identify and analyse the adverse effects of NT.

With government funds, some new institutions at universities already have been established to do research on these effects. Much of the research on effects is cross-disciplinary—a major handicap in enlisting the discipline-oriented universities to undertake the research. It also is a handicap in enlisting individual researchers who may not get academic rewards for working on interdisciplinary, applied problems. The question of how to create incentives for the private sector to conduct testing and do research on NT adverse effects is a different kind of institutional challenge.

Holding manufacturers responsible for the adverse effects of their products is clearly a major incentive for such testing and research. There is broad agreement that current government spending to determine the health and safety effects of NT is inadequate. At a Nov. 17, 2005, hearing before the House Committee on Science, industry, environmental and research organiza-tions called for federal spending of at least $100 million annually—more than three times current expenditures—on health and safety research. Some witnesses also called for a more actively managed, strategically targeted, and carefully coordinated approach to determine what adverse effects, if any, NT may create.

Recently, the Project on Emerging Nanotechnologies at the Woodrow Wilson Center released a comprehensive inventory of NT-related environmental, health and safety research. The inventory covered information on 208 research projects, in 6 countries and regions, accounting for $38 million of research annually. 169 projects, accounting for $27 million, were in the United States. However, within the United States, only $6 million of federal research was considered to be of "high relevance" to determining the environmental, health and safety implications of engineered nanomaterials.

PUBLIC PARTICIPATION

The involvement of members of the general public is crucial for dealing with NT's adverse effects in two different ways. First, as the Royal Society states, "some of the social and ethical concerns that certain applications of nanotechnologies are likely to raise stretch well beyond the basic science or engineering of the matter". Social values, apart from scientific questions, are an inextricable part of assessing risks and the public needs to be involved in assessing NT's risks, as well as in defining the measures to be taken to deal with the risks. The social value questions that NT is likely to raise transcend the risks of individual technologies or applications.

A study comparing the controversy over biotechnology with what might be expected in relation to NT noted that the issues many people saw as being involved in the biotechnology controversy included "Global drives towards new forms of proprietary knowledge; shifting patterns of ownership and control in the food chain; issues of corporate responsibility and corporate closeness to governments; intensifying relationships of science and scientists to the worlds of power and commerce; unease about hubristic approaches to limits in human understanding; conflicting interpretations of what might be meant by sustainable development…".

As the authors point out, these kinds of concerns cannot be accommodated within a framework of risk assessment of individual NT products. The second way the public needs to be involved is as consumers both of information about NT and of NT products. The tremendous potential of the technology will not develop without the public having a realistic view of the intended and unintended effects of NT. In its scope and diversity of applications, NT often has been compared with biotechnology. The lessons learned from biotechnology about the consequences of a poorly informed public should not be lost on those dealing with NT.

Some institutions have risen to this challenge. For example, DuPont has joined with the non-governmental organization Environmental Defence to define a process for identifying and reducing potential health, safety and environmental risks of nanoscale materials, and to test the process on specific DuPont materials or applications. A study by the Woodrow Wilson Center's Project on Emerging Nanotechnologies showed more trust in regulatory agencies than might have been expected. Almost half the sample believed that EPA, OSHA and other regulatory agencies would effectively manage risks from nanotechnology, while about 40 per cent believed the agencies could not be trusted.

Of the 177 participants in the study, 55 per cent said that voluntary standards applied by industry would not be sufficient to deal with NT risks. When asked for their top choice of how government and industry could increase public trust, 34 per cent selected increased safety tests before products are marketed, and 25 per cent chose supplying more information to support informed consumer choices. New and better institutions for public participation are needed. Going through the motions of a public hearing where nobody listens, or launching a slick public relations campaign will not suffice. The public needs to be educated, not brainwashed. It needs to be seriously listened to, not tolerated. And there needs to be real participation. New forms of participation are being tried, for example, using the Internet, and NT provides a good opportunity to use and experiment with these new methods.

HEALTH AND ECOLOGY

In a witty review of popular attitudes towards food, David Cort remarks that he has encountered absurdly conflicting dietary recommendations in books and articles on nutrition. One authority advises his readers to avoid fat; another advises that it should be eaten, on the theory that fat is the least fattening of all foods. Potatoes have been alternately damned and praised; mixed meals have been both frowned upon and approved; the consumption of alcohol, criticized and recommended. "The cruelty and irresponsibility of offering one single, standard dietary solution for everybody is obvious," Cort adds.

"What a person eats is, in many important respects, his life. But every individual is different from every other individual. No doctrinaire solution will work for them all. Each individual has conditioned habits of eating, of taste, of appetite, of expenditure of energy, of nervous rhythms and, most important of all, of metabolism. Few modern doctors have the interest, time or genius to find out all about any one individual; and many are themselves overweight. But the individual has the time and interest and, at least about himself, perhaps the genius. After all, it's his life."

Although Cort's remarks are confined to nutrition, they are quite relevant to a number of problems in human ecology. It would be an error to form rigid concepts of a normal man, a normal diet, or a normal way of life. Any such image would be woefully lacking in biochemical and physiological support. The biologist and physician must still admit, as Carrel did nearly a quarter of a century ago, that "most of the questions put to themselves by those who study human beings remain without answer. Immense regions of our inner world are still unknown." Impressive advances, to be sure, have been made during the past two decades in solving major problems of cellular biochemistry, endocrinology and physiology. Knowledge of the transformation of energy in the cell is very far advanced, for example and great strides have been made in understanding the chemistry of the genetic material in the cell's nucleus.

Despite this newly acquired knowledge, however, nutritionists have an extremely limited understanding of the function of many trace elements in the body and of the way in which vitamins prevent deficiency diseases. The nutritional requirements of the various tissues, essential details of the blood-forming system, the mechanisms of resistance to disease and many key relationships within the body await clearer understanding.

The human organism is extremely sensitive to very small quantities of hormones and nutrients. A variation by only a few drops in the daily output of insulin, a pancreatic hormone, will determine whether an individual will handle sugar properly or succumb to diabetes. A difference of a few thousandths of a gram (milligrams) in the intake of certain vitamins and trace elements makes for either well-being or nutritional disorders. According to the National Research Council, a man twentyfive years of age requires a daily intake of 1.6 milligrams of thiamine and riboflavin, 7.5 milligrams of vitamin C and 12 milligrams of iron to preserve good health.

His minimum daily requirements of iodine and copper are believed to be as little as 0.1 and 0.2 milligrams, respectively. If the daily intake of these nutrients is reduced by a few thousandths of a gram or less for a long period of time, the result will be a marked impairment of health. Despite the minute quantities involved, deficiencies of vitamins and minerals occur on a fairly large scale in the United States. If A. F. Morgan and L. M. Odland are correct in their claim that a dry, roughened skin and lesions of the eyes and mouth "may be considered a useful pointer to possible dietary faults," about 50 per cent of the adolescents examined in the affluent northeastern states and 20 per cent in the western states may have nutritional disorders.

We tend to look upon diseases as though they were localized disturbances, confined to definite parts of the body. Although it is customary to acknowledge the body's unity in sickness and in health, the whole is often overlooked in favour of the parts and vital interrelationships are ignored in favour of isolated events. Actually, every major deficiency and physical impairment has far-reaching effects. "Disease consists of a functional and structural disorder," Carrel notes. "Its aspects are as numerous as our organic activities. There are diseases of the stomach, of the heart, of the nervous system, etc. But in illness the body preserves the same unity as in health. It is sick as a whole. No disturbance remains strictly confined to a single organ. Physicians have been led to consider each disease as a specialty by the old anatomical conception of the human being. Only those who know man both in his parts and in his entirety, simultaneously under his anatomical, physiological and mental aspects, are capable of understanding him when he is sick."

A clear understanding of illness and health requires a greater appreciation not only of man but of men. Physical needs and responses vary enormously from one individual to another. Roger J. Williams, one of America's most eminent biochemists, emphasizes that within the same racial groups, even in the same families, men differ from one another in the shape and weight of their vital organs, the composition of body fluids, the output of endocrine glands, the rate of metabolism, temperature control and many other respects.

Individual differences in the amount of nutrients required for health and in the amount of nutrients absorbed by the body are of major practical importance. Individual requirements of calcium, trace elements, amino acids and vitamins are highly variable. For example, two normal five-year-old children who ate the same food and were exposed to the same environmental conditions showed a great variation in the amount of calcium they retained from their diets.

One child retained 264 milligrams of calcium per day for 45 days, whereas the other retained 469 milligrams, a difference of nearly 80 per cent. A high degree of variability is also found in individual responses to common environmental toxicants. Williams cites an account of 78 men who were exposed to vapors of carbon tetrachloride, of whom 15 suffered acute poisoning. Six became sufficiently ill to require hospitalization. "All were white men within 5 to 8 years of the same age and in general good health," observes Fredrich H. Harris, the source for Williams's account. "All 15 of the men poisoned were exposed from 3 to 8 hours; however, many of those who developed no symptoms were exposed for a similar length of time or longer."

The fact that man's knowledge of important bodily processes is incomplete and that many individuals exhibit marked deviations from "normal" responses and needs does not imply that the role of medicine is inconsequential in promoting health or that scientific generalizations are useless. The healthiest of men will undoubtedly require medical attention at some time in their lives and every physician must use physiological and anatomical generalizations as points of departure for the diagnosis of a disorder. Basic research in biochemistry and physiology promises to increase vastly our understanding of the preconditions for human health and well-being. But a phenomenon does not cease to exist because it cannot be readily explained by current methods and theories.

On the contrary, we are often profoundly affected by phenomena that are too complex to lend themselves to precise analysis at the present stage of scientific development.

The way in which stress produces illness and the role that nutrition plays in the occurrence and prevention of chronic diseases constitute fields of research that have scarcely been penetrated. It would be a major setback if the roles of these factors were minimized simply because they present formidable problems to research. Man must try to conserve many practices that, on the basis of long experience, are known to promote good health, even if the reasons why they are beneficial cannot be stated in precise biochemical terms.

What are some of these practices? As it is clear that our bodily functions are interdependent, we should not permit any part of our biological equipment to atrophy. This means that we must use all the parts of our bodies. Our musculature must be as fully employed as our minds. Many activities and experiences that have been restricted by our sheltered civilization seem to play important roles in preserving health. There is a suspicion that modern man indirectly harms his body by failing to use certain physiological resources that once supported his ancestors during periods of hunger and cold.

"There exists in the body of man, as of all animals, biological mechanisms for the storage of food developed for meeting the irregularities and cyclical changes in nature," Rene Dubos writes. "It may still turn out that a nutritional way of living permitting continuous growth at a maximum rate may have unfortunate distant results. Fasting fads may have some justification after all by providing an opportunity for the operation of certain emergency mechanisms built by nature into the human body."

Although it remains to be seen whether this conclusion will be accepted by nutritionists and physicians, it is reasonably clear that a one-sided manner of life is biologically undesirable. An existence anchored in either muscular exertion or inactivity, surfeit or poverty, stress or an overly sheltered manner of life, tends to produce physiological disequilibrium. A prolonged period of muscular exertion causes marked damage, whereas protracted physical inactivity appears to increase both the effect and the incidence of heart and circulatory disorders. When the coronary arteries of dogs are surgically narrowed to simulate atherosclerosis, a lack of exercise makes it difficult for new blood vessels to form and provide more blood to nourish the diseased area of the heart muscle.

There also seems to be more coronary illness among men whose occupations are sedentary than among those who engage in physical work. In contrast to the well-exercised individual, the "physically inactive individual shows signs of aging earlier in life," emphasizes Hans Kraus, of New York University. "He exists physiologically at a lower potential and is less well-equipped to maintain homeostasis [stable bodily conditions] and to meet daily stresses. This low level of function, combined with enforced suppression of the 'fight and flight' response, enhances the incidence of disease."

Not only does health require all the elements of a full life; it also requires that they exist in balance. A moderate amount of exertion, a balanced intake of nutriment and a reasonable amount of exposure to stress are necessary for the maintenance of well-being.

Almost any kind of excess is harmful. This has become apparent in respect to nutrition. "In laboratory experiments rats fed an unlimited diet were found to die sooner than animals prevented from gaining weight by a diet severely restricted in quantity but well balanced in composition," Dubos observes. "Likewise insurance statisticians have repeatedly emphasized that in man the obese have a short expectancy of life. Indeed, obesity is now publicized as the most common nutritional disease of Western society. This was apparently true also in Imperial Rome, as during all periods of great material prosperity.

'In the old days,' wrote Lucretius in the fifth book of De Rerum Natura, 'lack of food gave languishing limbs to Lethe; contrariwise today surfeit of things stifles us.' Thus, history repeats itself. Like the prosperous Romans of two thousand years ago, countless men of the Western world today are digging their own graves through over-eating."

Finally, if man is to maintain good health, he requires diversity. It is hardly necessary to emphasize that a healthy psyche and a rounded response to stress are nourished by a variety of experiences and by diversified surroundings. However, it does need to be emphasized that variety is also necessary for the satisfaction of man's nutritional needs. By limiting his selection of foods, he may fail to acquire nutrients known to exist in natural foods but not yet identified by chemists or nutritionists. "Because we know that unidentified factors exist in foods of plant and animal origin," observes George M. Briggs, of the National Institutes of Health, "it is wisest to eat a wide variety of foods from the many excellent food groups."

If this is done regularly, Briggs emphasizes, it is unnecessary to supplement the diet with vitamins and so-called health foods. It might be added that if food such as grain were not overly processed, it would be unnecessary to "enrich" bread with synthetic vitamins, which replace only part of the valuable nutrients lost in the milling process. In general, the more a food is processed, the greater the likelihood that identified nutrients as well as unidentified ones will be removed. Completeness, balance and diversity should be regarded as practical ecological concepts—as important in producing healthy human communities as they are in producing stable plant-animal communities. Indeed, it is not farfetched to say that when these concepts are correctly applied, they promote human health because they produce a stable ecosystem of men, animals and plants. In the last analysis, the ecology of health is grounded in natural ecology. A complete way of life for man presupposes unrestricted access to the countryside as well as the town, to soil as well as to pavement, to flora and fauna as well as to libraries and theaters. A balanced way of life presupposes a lasting equilibrium between land and city; animals, men and plants; air, water and industry.

Diversity presupposes an awareness that nearly every species perpetuates the stability of the biosphere, either directly or indirectly. As the city encroaches on the land, however, human life becomes increasingly restricted to a polluted, nerve-racking urban environment. As the city begins to dominate the land, physical activity yields to sedentary forms of work. Finally, as the aggregations of population swell to massive proportions, foods become laden with chemicals and human activities are standardized. By oversimplifying the natural environment, we have created an incomplete man who lives an unbalanced life in a standardized world. Such a man is ill—not only morally and psychologically, but physically.

ECOLOGY AND HUMAN BEHAVIOUR

Disease is an inevitable part of life and coping with disease is a universal aspect of the human experience. All humans, during the course of their lives, harbour infections by disease organisms and suffer the consequences of those infections. The experience of disease, by individuals or whole populations, is as inescapable as death itself. Yet the particular diseases that afflict people, as well as the way in which symptoms are interpreted and acted upon, vary greatly by culture. Understanding the nature of interactions between disease and culture can be a productive way of understanding humanity and is therefore an important topic in medical anthropology. Because of its biocultural, evolutionary and crosscultural perspectives, anthropology has much to offer to the understanding of the causes and consequences of disease. From an anthropological perspective, diseases cannot be explained as purely "things in themselves"; they must be analyzed and understood within a human context—that is, in relation to ecology and culture.

The distribution of disease in a population is neither constant nor random. The diseases that characteristically afflict members of a population vary significantly among societies because of differences in culture, ecological setting and historical period. More important, within a single society, there may be striking variations in the kind and severity of diseases that afflict individuals of different ages, sexes, social classes and ethnic groups. Understanding the descriptive epidemiological distribution of disease morbidity and mortality is important to medical anthropologists particularly because the social patterning of disease distribution often reflects the cultural coding of behaviours. Understanding the role of culture in disease distribution is also necessary for the implementation of successful disease control programmes

Culture plays a major role in determining the patterns of disease and death in a population for two reasons. First, culture may shape important behaviours (with respect to diet, activity patterns, water use, sexual practices, etc.) that predispose individuals to acquire certain diseases. Second, through culture, people actively change the nature of their environment, often in ways that affect their health. The archaeological and historical record clearly demonstrates that the environmental changes caused by humans can have profound effects, both positive and negative, on disease rates. Although humans have a dual system of inheritance through both genes and culture, culture is the primary mechanism for survival. Culture is a mechanism of adaptation to environmental threats, such as diseases, which act as agents of natural selection in the evolution of both human biology and culture. Cultural practices, however, can also be maladaptive when they exacerbate health problems.

Ecology is the study of the relationship of organisms in an environment. Human societies coinhabit their environment with many other organisms, including those producing disease. An ecological approach to human health and illness emphasizes the fact that the environment and its health risks are, to a significant extent, created by the culture. In many cultures, people think of themselves as masters of their environment, because they exploit so many plants and animals within the food chain as sources of energy and nutrients. Yet at the same time, humans are being exploited by microorganisms, including those that cause disease, as a source of food and shelter. Disease ecology primarily focuses on the multiplex interactions of the pathogen, the environment and the human host. The nutritional, physiological, genetic and mental condition of the human individual host all play significant, if still incompletely understood, roles in both infectious and non-infectious disease states. The current international health problems called "emerging infections" provide a good example of the mutability of these ecologic interactions.

The study of disease and human behaviour in an ecological setting is a fundamental task for medical anthropology. The approach contributes to basic and applied research in the field by providing a strategy for answering some of the major questions raised by both general anthropology and epidemiology. For example, it can be applied to anthropological questions concerning the interaction of biology and culture in human evolution or to questions of why particular cultural behaviours may "make sense" and be retained in an ecological setting.

This type of research strategy is truly biocultural and can therefore help to bridge the gap between biological and cultural anthropology. In epidemiology, the contribution of an anthropological focus on human behavioural patterns can help unravel fundamental questions of disease causality. The study of disease, ecology and behaviour also has important implications for public health programmes. Through the study of behavioural patterns related to the social epidemiology of disease, it is possible to design health programmes that are both effective and culturally acceptable. Moreover, the ecological approach can help to anticipate the health implications of technological change or new political-economic policies.

This chapter discusses basic concepts and methods in the study of disease, ecology and behaviour; it is not an exhaustive review of the literature. Instead, we summarize many illustrative examples of research that focuses on particular diseases in particular cultural settings. These examples emphasize infectious diseases for several reasons: first, the etiology (causation) of these diseases is the best understood; second, the ecological approach described here is most applicable to this category of disease, (although it is also applicable to the study of chronic diseases with complex etiologies, which are characteristic of industrialized societies, as well as nutritional deficiency diseases, which are characteristic of the poor in economically developing societies); and, finally, infectious diseases still represent the major cause of morbidity and mortality in non-industrialized societies, which anthropologists have traditionally studied.

DISEASE AND HUMAN BEHAVIOURAL PATTERNS

As the field of epidemiology has made clear since its inception in the late 1800s, diseases are not distributed randomly in human populations. Some individuals—and some groups of individuals—are at increased risk from various diseases, for masons that are often unclear. Epidemiologists not only describe patterns of disease occurrence through space and through time, but they attempt to elucidate disease etiology through the search for risk factors that appear to be significantly associated with disease outcome. Disease risk factors are of two major types. Endogenous risk factors are those that are biologically intrinsic to the human host. For example, genetic diseases, such as sickle cell anemia or hemophilia, have, by definition, an endogenous etiology. More commonly, however, genetic inheritance implies a predisposition to a disease that requires other variables or cofactors for expression to occur.

Exogenous risk factors are those that am extrinsic to the body of the human host. Some of these may be biotic, such as microorganisms that cause infectious diseases; others are non-biotic substances present in the environment, such as toxic chemicals in the workplace. In most cases of disease, both endogenous and exogenous factors are involved—hence, the notion of "multiple causation," or "multifactorial etiology". Humans may unwittingly increase the likelihood of disease by exposing themselves or others to risk factors of both the exogenous and endogenous variety. In many cases, this enhanced exposure potential occurs through disruption of existing ecological relationships between the host, the agent(s) of disease and the environment. In this way, human behaviour itself may be said to be a risk factor for disease, in that human activity may be a necessary component in the chain of events leading to a disease outcome.

Anthropologists, as professional observers and interpreters of human behaviour, have an obvious and crucial role to play in the understanding of disease etiology: they can facilitate risk factor identification by describing distinctive patterns of human behaviour related to the social distribution of disease. In this capacity, anthropologists may contribute directly to the generation of causal hypotheses, as they did in the case of kuru and cannibalism in New Guinea. In addition, anthropological descriptions of risk factor exposure based on long-term ethnographic observation may be more valid than those normally obtained through the standard epidemiological technique of questionnaire surveys.

Perhaps most important, anthropologists are especially equipped to understand disease-promoting human behaviours in sociocultural context. This includes the distribution of these behaviours through space and time, as well as the ideological and political-economic factors that serve to legitimate these behaviours. It is in this latter capacity—as interpreters of human behaviour who elucidate how and why people act the way they do—that anthropologists may contribute directly to medical anthropological theory building and indirectly to disease prevention and control.

PUBLIC ROLE IN HEALTH CARE MATTERS

Health care in about 1990 cost at least $1.7 trillion, or about 8 per cent of world income, making it one of the largest industries in the global economy. On average, 60 per cent of this is public spending. If this spending is excessive or otherwise inappropriate, the consequences for the economy and for health outcomes could be substantial. Governments also provide a large share of health services, sometimes as large as the share in spending, and often intervene in various ways in the private health care market. Since most health care is a private good, it is surprising that so much of it is provided, financed or regulated by the state. In contrast to what happens in many other sectors of the economy, this substantial public role is most pronounced in high income countries which are generally very market-oriented; the state usually finances a smaller share of health care in poorer countries.

Does this pattern provide models for the future development of the health sector and the public role in it, for lowincome countries? And is the variation among countries in the amount and kind of public intervention associated with differences in people's health, in what is spent on health care or in how well health systems function?

Besides consuming large resources, many health systems are regarded as inefficient or inequitable or both; they are often described as in "crisis", as needing "reform", or as having "failed". Are the supposed failures of health systems real? If so, they might be caused by misguided public intervention, so they could be corrected by a smaller or different public role and a greater reliance on private markets. Or governments might intervene for sound reasons, to correct or compensate for failings in those markets; that is, outcomes would be even worse if left entirely to the private sector. Is there an appropriate frontier between private and public action, and a best combination of instruments for the state to use when it intervenes? These and related questions must be confronted in any reform effort. It is relatively simple to conclude that governments should do certain things and should leave others to private activity; but often there is a variety of possible solutions and no obviously best approach. Theory does not always provide clear answers, and the empirical evidence is incomplete, extremely varied and difficult to interpret.

CHOICES FOR STATE INTERVENTION

It matters not only *whether* governments intervene, but also *how* they do it: the second essential question is what the public sector should do, given that some problem in the private market appears to warrant some public action. This is particularly important because government failings in the health sector are also common, and often result from intervening in the wrong ways or with the wrong instruments.

There are five distinct instruments of public intervention: arranged from the least to the greatest intrusion into private decisions, these are to:

1. *Inform,* which may mean to persuade, but does not require anyone to do anything. Governments do this when they publicize the health risks of smoking, or include health and basic hygiene education in public schools. These are examples of information directed at consumers, but governments also inform health care providers and suppliers of health care inputs, as by conducting research and disseminating information on disease patterns and on the effects and risks of medical procedures.
2. *Regulate,* which determines how a private activity may be undertaken. Governments sometimes regulate the medical profession by setting standards for doctors or accrediting hospitals, although these activities may also be undertaken by private bodies. And government regulation is common in the insurance industry, in the importation of medical equipment, drugs and supplies and in the protection of food and water quality. More generally, governments can influence private health care activity in many ways, often combining regulation with some financial incentives to offset the costs but without public financing. Regulation is usually pursuant to a law, and is often determined by an executive or administrative body.
3. *Mandate,* which obligates someone to do something and to pay for it. Compliance with regulations can also imply substantial private costs; but a mandated activity is different in that it *must* be performed, whereas a private producer can react to regulation by choosing not to undertake the activity. Mandates are usually specified in law, which may subsequently be adumbrated by regulation. The most important mandates, in financial terms, are the requirements that employers provide health services or insurance to their employees, or contribute to social insurance funds for that purpose. Governments can also impose mandates on individuals, as by requiring that children entering school be immunized.
4. *Finance* health care with public funds. Because mandated insurance is effectively paid for by an earmarked, involuntary contribution which is equivalent to a tax, "public" health expenditure is commonly defined to include such costs along with expenditures from public budgets. The obverse of spending public funds is to tax particular activities or goods, such as alcohol and tobacco, at least partly for health reasons. This issue is not treated here; while taxation which reduces consumption of specific

goods may have substantial health effects, it is always limited to very few goods and is not used systematically to promote health. Finally, the state may—

5. *Provide* or deliver services, using publicly-owned facilities and civil service staff. This is what Ministries of Health in most poor countries do; so do various governmental bodies in many countries at all income levels. Once a society has decided to finance health services with public funds, the choice arises of whether to provide them through public facilities or to pay private producers to provide them. The appropriate way to consider this choice is as a standard "make or buy" decision. The issues for a government are the same as for a private firm, and turn on costs—is it cheaper to produce something than to buy from an outside supplier?—and on the risks and difficulties of enforcing contracts and avoiding fraud when dealing with such suppliers.

Because public financing requires public resources, it is perhaps the crucial choice about state action. However, all the instruments mentioned have costs; even information is not free. The benefits from any intervention have always to be weighed against these costs.

In addition, sometimes two instruments overlap, or one requires the use of another: mandates imply regulation, and public provision usually implies at least partial public finance. And two instruments can be alternatives: for part of the population, governments can either finance health care or mandate financing by employers or other private institutions. Differences in ability to pay make it natural to operate mixed systems of public intervention, such as mandated coverage for the non-poor and public finance for the poor. There are alternatives even within mandated coverage, such as "play or pay" arrangements in which employers can finance health care directly for their employees or pay into a social security scheme.

THE APPROPRIATE PUBLIC ROLE IN HEALTH

As a first approximation, it is easier to say what govern-ments *should not do* in health than to specify what they *should do.* That is, it is clear that certain actions are likely or certain to violate one or more of the objectives of a health care system to an important degree. To apply a topographic metaphor: such actions correspond to falling off a plateau of satisfactory outcomes, all somewhat different but none clearly dominating the rest, and into one of the surrounding chasms of one or another kind of failure. Four "don'ts" are discussed in what follows: they refer to the way the public is taxed, charged or exempted to pay for health care; the providers to which governments transfer public funds; the way providers are paid; and the services they are paid to provide.

WHAT GOVERNMENT SHOULD NOT DO

The first thing governments should not do is to use the tax system, or any system of fees at public facilities, to make the poor subsidize the health care of the rich. Conceptually, subsidies are justified only for the poor, and broader financing of care through insurance is a question of efficiency rather than of equity. This is not only a matter of whether the rich use more of publicly-financed services than the poor do, although great inequities often arise because the rich have more access to those services. Financial equity also depends on who pays the taxes. More narrowly, governments should not contribute to social security financing from general revenues, unless coverage is universal, because when only part of the population is covered it is usually the poor who are excluded. And governments should not treat private insurance coverage as a cost to employers unless it is also treated as income to beneficiaries. Such practices are not only inequitable; they are also inefficient to the extent that they lead to excessive spending on health care, or reduce labour mobility.

Controlling inequitable subsidies does not mean that social security systems should be dismantled or must be made universal. Even when incomplete, such mandated insurance includes substantial progressive subsidies from high-paid to low-paid workers, and the mere fact that some people in society receive more generous health care is not necessarily a problem, so long as they pay for it. What matters is that governments not make everyone pay for what some are excluded *ex ante* from receiving. Perverse subsidies not only cause immediate inefficiencies or inequities, they also create interests that oppose subsequent health system reform. This is evident in the United States and Chile and it is the reason why the design of subsidies is a crucial part of any reform to extend or improve coverage for the poor. The second thing governments should not do is tie public finance to public provision. The choice of whether to provide care through public or parastatal facilities should be treated as a "make or buy" decision, subordinate to the larger decision about what to pay for. That does not necessarily mean eliminating public provision, which will sometimes be the best solution. It means rather that competition between public and private providers should be based on costs and on quality, and not on price to the consumer, as is commonly the case.

To achieve those goals, however, requires other changes in how public institutions operate, changes that probably cannot occur so long as public funds go automatically and exclusively to government facilities. This conclusion is most pertinent to poor countries where public systems are most likely to be inefficient and to be used only because they are free or nearly so. High-income countries with a large share of public provision, particularly in hospitals, suffer less from these problems.

A third thing governments should not do is pay for health care by fee-for-service, unless other mechanisms are used to control expenditures. This is seldom a problem with publicly provided services, but makes it hard to control costs when governments finance private providers. Even negotiating or controlling the fees is not enough, as the Canadian experience demonstrates, since providers can respond to fees they consider too low, by increasing output. This helps explain why Canadian health expenditure has risen faster than that in European countries which rely on other payment mechanisms.

European countries which pay for health care by fee-for-service also rely on global budgets, utilization reviews or other instruments of cost containment. Since expenditures equal prices times quantities, and since quantities of services respond to prices and are difficult to control directly, countries which pursue both macroeconomic and microeconomic efficiency in health spending usually control both prices and expenditures. In contrast to European and North American experience, Brazil applies this system to block grants for federal expenditure in states with relative financial autonomy; but in states where federal money is paid directly to providers, the federal government also controls one crucial quantity, the number of hospitalizations that federal money will pay for. In all cases, the government sets prices for services.

It also helps, in controlling expenditures, to define "services" as complete treatments for specific conditions rather than as all the individual components of such treatments, so they can be financed by mechanisms such as the Diagnostic Related Group payments used in the United States and the provider can be required to assume part of the financial risk. Changing from a system based on overall budgets and on salaries for providers, to a fee-for-service system without offsetting the resulting incentives to overprovision can lead to an explosion of costs, as in the Czech Republic. Finally, if governments mean to pursue some combination of better health and lower costs, they should not—in fact they cannot—simply finance whatever people demand when care is free to consumers.

This does not mean governments should not "subsidize demand" rather than "subsidizing supply" or providing services. It means only that there must be limitations on what will be paid for publicly. Such limitations are a particularly contentious matter, because both good health and cost containment may be opposed to consumer or provider satisfaction, which are also politically important objectives.

Nonetheless, two empirical observations are germane to this decision. One is that private insurance always carries some limitations, either as to the services covered or as to cost-sharing. Except where poverty is important, so that cost-sharing is more difficult, there is no reason for public finance to be systematically more generous in this regard than private risk-sharing arrangements. The second observation is that private markets will tend to supply what people demand, and public intervention typically acts to emphasize needs instead. In other forms of public subsidy, it is common to distinguish between wants and needs, and concentrate spending on the latter. This is the case, for example, with food subsidies, which are also very much healthrelated. Price subsidies are usually limited to "basic" foodstuffs, and food stamps cannot legally be used to buy alcohol or other non-necessities.

This may be considered partly a matter of efficiency, assuring more health gain than would otherwise occur. But it is also a matter of equity: in contrast to actuarial private insurance, where every purchaser buys the expected value of the health services needed, public finance is involuntary. It comes from taxpayers who have a legitimate interest in meeting needs, and thereby getting value for their money, but not necessarily in paying for wants.

None of these "don'ts" is easy to implement, because some consumer or provider interests can be expected to oppose every one of them, in every country. But they are arguably the most important conclusions about where the public/private frontier should run in the health sector. It is notable that none of these conclusions depends solely on features unique to the health sector. The peculiarities of health matter most for the difficulty of distinguishing "needs" from wants, and—partly as a consequence—for the dangers of paying providers for whatever they choose to provide, without incentives to control costs. A system which avoided the problems would still present complex and difficult questions for the proper role of the state, but it would have more latitude to pursue improvements. And it would not matter so much, exactly what objectives the government pursued or which combination of instruments it applied.

WHAT GOVERNMENT SHOULD DO

Beyond the prescriptions for how the state should deal with the problems of each of the three domains of health care, several "dos" appear to be generally valid for governments. If the objective is to minimize deadweight losses from public intervention and leave as much room as possible for private choices, then the first thing governments should do is to use each less-intrusive instrument to the point where a more intrusive intervention is justified, following the sequence of increasingly greater interference— inform, regulate, mandate, finance and provide services. That is, governments should regulate private activity when merely improving people's information is not enough, deliver services if it is infeasible to finance private providers equitably, and so on. Public finance is inescapable for some actions, but particularly in low- and middle-income countries much can probably be accomplished by better use of information and regulation. Failure to use these other instruments well, can increase the need for public finance.

Sometimes the problem is that governments exploit these instruments too little. They do not regulate private insurance when it first begins to expand, which makes subsequent regulation politically more difficult; or they do not initially react when health-damaging behaviours such as smoking become more entrenched. Sometimes the problem is inappropriate regulation, which needlessly restricts competition, or enforces inefficiency in the public sector by centralizing nearly all decisions. And poor countries in particular often get the worst of both worlds by paying for activities which should be, but are not, regulated: government subsidy of medical education without adequate control of quality or relation to needs is a common example.

In the worst of cases, governments use all the available instruments in exactly the reverse order. They try to provide more health care than they can pay for, with the result that most services are under-financed and of poor quality; they try to finance services, some of which might be mandated and paid for by consumers or employers; they mandate care, as by social security systems, without adequately regulating it; and they do too little to inform the public and providers either of dangers to health or of how the health care system is actually working.

Much of the criticism of government failure in the health sector, especially in poor countries, describes the result of getting things backwards. These ideas are represented in Figure, which shows the appropriate relation among the instruments by which the state can intervene in health care which portrays the kind of inappropriate or imbalanced relation often found in poor nations.

Figure indicates that whatever is mandated, financed or provided publicly should also be regulated, and much else besides. The state can finance some care that is not anyone's mandated responsibility, and can also mandate care or insurance coverage which is financed privately. And most if not all public provision should be fully financed by government, but possibly with much more scope for finance than for provision by the state.

5

Nanotechnology in Environmental Pollution and Treatment

NANOPOLLUTION

Ecotoxicological impacts of nanoparticles and the potential for bioaccumulation in plants and microorganisms remain under-researched. The capacity for nanoparticles to function as a transport mechanism also raises concern about the transport of heavy metals and other environmental contaminants. A May 2007 Report to the UK Department for Environment, Food and Rural Affairs noted concerns about the toxicological impacts of nanoparticles in relation to both hazard and exposure. The report recommended comprehensive toxicological testing and independent performance tests of fuel additives. Not enough data exists to know for sure if nanoparticles could have undesirable effects on the environment. Two areas are relevant here:

- In free form nanoparticles can be released in the air or water during production (or production accidents) or as waste by-product of production, and ultimately accumulate in the soil, water or plant life.
- In fixed form, where they are part of a manufactured substance or product, they will ultimately have to be recycled or disposed of as waste.

It is not known yet whether certain nanoparticles will constitute a completely new class of non-biodegradable pollutant. In case they do, it is not known how such pollutants could be removed from air or water because most traditional filters are not suitable for such tasks (their pores are too big to catch nanoparticles).

Of the US$710 million spent in 2002 by the U.S. government on nanotechnology research, only $500,000 was spent on environmental impact assessments. Risks identified by Uskokovic (2007) include: self-replicating nanobots aggressively or through slowly rising supremacy wiping out the whole biosphere; further destabilising the already endangered diversity of the biosphere. Concerns have been raised about Silver Nano technology used by Samsung in a range of appliances such as washing machines and air purifiers.

Life Cycle Responsibility

To properly assess the health hazards of engineered nanoparticles the whole life cycle of these particles needs to be evaluated, including their fabrication, storage and distribution, application and potential abuse, and disposal. The impact on humans or the environment may vary at different stages of the life cycle.

The Royal Society report identified a risk of nanoparticles or nanotubes being released during disposal, destruction and recycling, and recommended that "manufacturers of products that fall under extended producer responsibility regimes such as end-of-life regulations publish procedures outlining how these materials will be managed to minimize possible human and environmental exposure".

Reflecting the challenges for ensuring responsible life cycle regulation, the Institute for Food and Agricultural Standards has proposed standards for nanotechnology research and development should be integrated across consumer, worker and environmental standards. They also propose that NGOs and other citizen groups play a meaningful role in the development of these standards.

FORMS OF POLLUTION

The major forms of pollution are listed below along with the particular pollutants relevant to each of them:

- Air pollution, the release of chemicals and particulates into the atmosphere. Common gaseous air pollutants include carbon monoxide, sulfur dioxide, chlorofluorocarbons (CFCs) and nitrogen oxides produced by industry and motor vehicles. Photochemical ozone and smog are created as nitrogen oxides and hydrocarbons react to sunlight. Particulate matter, or fine dust is characterized by their micrometre size PM10 to PM2.5.
- Water pollution, by the release of waste products and contaminants into surface run-off into river drainage systems, leaching into groundwater, liquid spills, wastewater discharges, eutrophication and littering.
- Soil contamination occurs when chemicals are released by spill or underground leakage. Among the most significant soil contaminants are hydrocarbons, heavy metals, MTBE, herbicides, pesticides and chlorinated hydrocarbons.
- Littering
- Radioactive contamination, resulting from 20th century activities in atomic physics, such as nuclear power generation and nuclear weapons research, manufacture and deployment.
- Noise pollution, which encompasses roadway noise, aircraft noise, industrial noise as well as high-intensity sonar.

- Light pollution, includes light trespass, over-illumination and astronomical interference.
- Visual pollution, which can refer to the presence of overhead power lines, motorway billboards, scarred landforms (as from strip mining), open storage of trash or municipal solid waste.
- Thermal pollution, is a temperature change in natural water bodies caused by human influence, such as use of water as coolant in a power plant.

Pollutants

A pollutant is a waste material that pollutes air, water or soil. Three factors determine the severity of a pollutant: its chemical nature, the concentration and the persistence

Sources and Causes

Air pollution comes from both natural and manmade sources. Though globally manmade pollutants from combustion, construction, mining, agriculture and warfare are increasingly significant in the air pollution equation.

Motor vehicle emissions are one of the leading causes of air pollution. China, United States, Russia, Mexico, and Japan are the world leaders in air pollution emissions. Principal stationary pollution sources include chemical plants, coal-fired power plants, oil refineries, petrochemical plants, nuclear waste disposal activity, incinerators, large livestock farms (dairy cows, pigs, poultry, etc.), PVC factories, metals production factories, plastics factories, and other heavy industry. Agricultural air pollution comes from contemporary practices which include clear felling and burning of natural vegetation as well as spraying of pesticides and herbicides

Some of the more common soil contaminants are chlorinated hydrocarbons (CFH), heavy metals (such as chromium, cadmium—found in rechargeable batteries, and lead—found in lead paint, aviation fuel and still in some countries, gasoline), MTBE, zinc, arsenic and benzene. In 2001 a series of press reports culminating in a book called Fateful Harvest unveiled a widespread practice of recycling industrial byproducts into fertilizer, resulting in the contamination of the soil with various metals.

Ordinary municipal landfills are the source of many chemical substances entering the soil environment (and often groundwater), emanating from the wide variety of refuse accepted, especially substances illegally discarded there, or from pre-1970 landfills that may have been subject to little control in the U.S. or EU. There have also been some unusual releases of polychlorinated dibenzodioxins, commonly called dioxins for simplicity, such as TCDD.

Pollution can also be the consequence of a natural disaster. For example, hurricanes often involve water contamination from sewage, and petrochemical spills from ruptured boats or automobiles. Larger scale and environmental damage is not uncommon when coastal oil rigs or refineries are involved. Some sources of pollution, such as nuclear power plants or oil tankers, can produce widespread and potentially hazardous releases when accidents occur.

In the case of noise pollution the dominant source class is the motor vehicle, producing about ninety per cent of all unwanted noise worldwide.

EFFECTS

Human Health

Adverse air quality can kill many organisms including humans. Ozone pollution can cause respiratory disease, cardiovascular disease, throat inflammation, chest pain, and congestion. Water pollution causes approximately 14,000 deaths per day, mostly due to contamination of drinking water by untreated sewage in developing countries.

Oil spills can cause skin irritations and rashes. Noise pollution induces hearing loss, high blood pressure, stress, and sleep disturbance. Mercury has been linked to developmental deficits in children and neurologic symptoms.

Lead and other heavy metals have been shown to cause neurological problems. Chemical and radioactive substances can cause cancer and as well as birth defects.

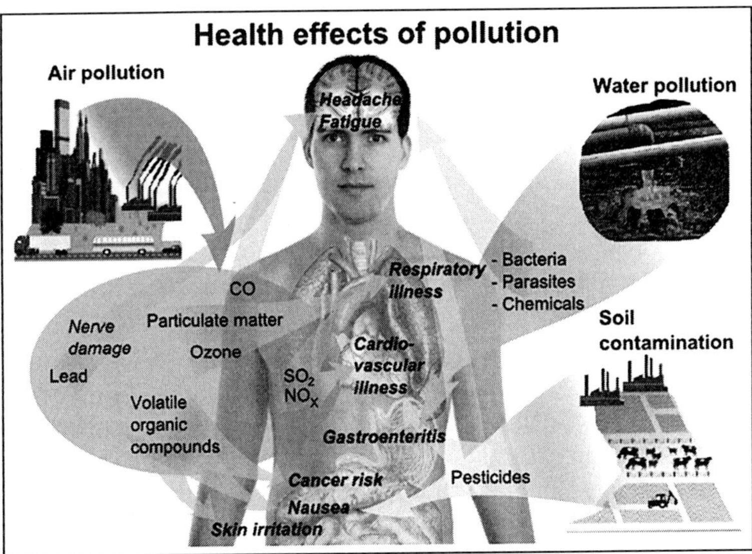

Ecosystems

- Sulphur dioxide and nitrogen oxides can cause acid rain which lowers the pH value of soil.
- Nitrogen oxides are removed from the air by rain and fertilise land which can change the species composition of ecosystems.

- Soil can become infertile and unsuitable for plants. This will affect other organisms in the food web.
- Smog and haze can reduce the amount of sunlight received by plants to carry out photosynthesis and leads to the production of tropospheric ozone which damages plants.
- Invasive species can out compete native species and reduce biodiversity. Invasive plants can contribute debris and biomolecules (allelopathy) that can alter soil and chemical compositions of an environment, often reducing native species competitiveness.
- Biomagnification describes situations where toxins (such as heavy metals) may pass through trophic levels, becoming exponentially more concentrated in the process.
- Carbon dioxide emissions cause ocean acidification, the ongoing decrease in the pH of the Earth's oceans as CO_2 becomes dissolved.
- The emission of greenhouse gases leads to global warming which affects ecosystems in many ways.

REGULATION AND MONITORING

To protect the environment from the adverse effects of pollution, many nations worldwide have enacted legislation to regulate various types of pollution as well as to mitigate the adverse effects of pollution.

Pollution Control

Pollution control is a term used in environmental management. It means the control of emissions and effluents into air, water or soil. Without pollution control, the waste products from consumption, heating, agriculture, mining, manufacturing, transportation and other human activities, whether they accumulate or disperse, will degrade the environment. In the hierarchy of controls, pollution prevention and waste minimization are more desirable than pollution control.

AIR POLLUTION

Air pollution is the introduction of chemicals, particulate matter, or biological materials that cause harm or discomfort to humans or other living organisms, or damages the natural environment, into the atmosphere.

The atmosphere is a complex, dynamic natural gaseous system that is essential to support life on planet Earth. Stratospheric ozone depletion due to air pollution has long been recognized as a threat to human health as well as to the Earth's ecosystems.

POLLUTANTS

An air pollutant is known as a substance in the air that can cause harm to humans and the environment. Pollutants can be in the form of solid particles, liquid droplets, or gases. In addition, they may be natural or man-made.

Pollutants can be classified as either primary or secondary. Usually, primary pollutants are substances directly emitted from a process, such as ash from a volcanic eruption, the carbon monoxide gas from a motor vehicle exhaust or sulfur dioxide released from factories.

Secondary pollutants are not emitted directly. Rather, they form in the air when primary pollutants react or interact. An important example of a secondary pollutant is ground level ozone - one of the many secondary pollutants that make up photochemical smog.

Note that some pollutants may be both primary and secondary: that is, they are both emitted directly and formed from other primary pollutants.

About 4 per cent of deaths in the United States can be attributed to air pollution, according to the Environmental Science Engineering Programme at the Harvard School of Public Health.

Major Primary Pollutants Produced by Human Activity Include

- Sulfur oxides (SO_x): Especially sulfur dioxide, a chemical compound with the formula SO_2. SO_2 is produced by volcanoes and in various industrial processes. Since coal and petroleum often contain sulfur compounds, their combustion generates sulfur dioxide. Further oxidation of SO_2, usually in the presence of a catalyst such as NO_2, forms H_2SO_4, and thus acid rain. This is one of the causes for concern over the environmental impact of the use of these fuels as power sources.
- Nitrogen oxides (NO_x): Especially nitrogen dioxide are emitted from high temperature combustion. Can be seen as the brown haze dome above or plume downwind of cities.Nitrogen dioxide is the chemical compound with the formula NO_2. It is one of the several nitrogen oxides. This reddish-brown toxic gas has a characteristic sharp, biting odour. NO_2 is one of the most prominent air pollutants.
- Carbon monoxide - is a colourless, odourless, non-irritating but very poisonous gas. It is a product by incomplete combustion of fuel such as natural gas, coal or wood. Vehicular exhaust is a major source of carbon monoxide.
- Carbon dioxide (CO_2): A greenhouse gas emitted from combustion but is also a gas vital to living organisms. It is a natural gas in the atmosphere.
- Volatile organic compounds: VOCs are an important outdoor air pollutant. In this field they are often divided into the separate categories of methane (CH_4) and non-methane (NMVOCs). Methane is an extremely efficient greenhouse gas which contributes to enhanced global warming. Other hydrocarbon VOCs are also significant greenhouse gases via their role in creating ozone and in

prolonging the life of methane in the atmosphere, although the effect varies depending on local air quality. Within the NMVOCs, the aromatic compounds benzene, toluene and xylene are suspected carcinogens and may lead to leukemia through prolonged exposure. 1,3-butadiene is another dangerous compound which is often associated with industrial uses.

- Particulate matter: Particulates, alternatively referred to as particulate matter (PM) or fine particles, are tiny particles of solid or liquid suspended in a gas. In contrast, aerosol refers to particles and the gas together. Sources of particulate matter can be man made or natural. Some particulates occur naturally, originating from volcanoes, dust storms, forest and grassland fires, living vegetation, and sea spray. Human activities, such as the burning of fossil fuels in vehicles, power plants and various industrial processes also generate significant amounts of aerosols. Averaged over the globe, anthropogenic aerosols—those made by human activities—currently account for about 10 per cent of the total amount of aerosols in our atmosphere. Increased levels of fine particles in the air are linked to health hazards such as heart disease, altered lung function and lung cancer.
- Toxic metals, such as lead, cadmium and copper.
- Chlorofluorocarbons (CFCs): harmful to the ozone layer emitted from products currently banned from use.
- Ammonia (NH_3): Emitted from agricultural processes. Ammonia is a compound with the formula NH_3. It is normally encountered as a gas with a characteristic pungent odour. Ammonia contributes significantly to the nutritional needs of terrestrial organisms by serving as a precursor to foodstuffs and fertilizers. Ammonia, either directly or indirectly, is also a building block for the synthesis of many pharmaceuticals. Although in wide use, ammonia is both caustic and hazardous.
- Odors: Such as from garbage, sewage, and industrial processes
- Radioactive pollutants: Produced by nuclear explosions, war explosives, and natural processes such as the radioactive decay of radon.

Secondary Pollutants Include

- Particulate matter formed from gaseous primary pollutants and compounds in photochemical smog.Smog is a kind of air pollution; the word "smog" is a portmanteau of smoke and fog. Classic smog results from large amounts of coal burning in an area caused by a mixture of smoke and sulfur dioxide. Modern smog does not usually come from coal but from vehicular and industrial emissions that are acted on in the atmosphere by sunlight to form secondary pollutants that also combine with the primary emissions to form photochemical smog.

- Ground level ozone (O_3) formed from NO_x and VOCs. Ozone (O_3) is a key constituent of the troposphere (it is also an important constituent of certain regions of the stratosphere commonly known as the Ozone layer). Photochemical and chemical reactions involving it drive many of the chemical processes that occur in the atmosphere by day and by night. At abnormally high concentrations brought about by human activities (largely the combustion of fossil fuel), it is a pollutant, and a constituent of smog.
- Peroxyacetyl nitrate (PAN) - similarly formed from NO_x and VOCs.

Minor Air Pollutants Include

- A large number of minor hazardous air pollutants. Some of these are regulated in USA under the Clean Air Act and in Europe under the Air Framework Directive.
- A variety of persistent organic pollutants, which can attach to particulate matter.

Persistent organic pollutants (POPs) are organic compounds that are resistant to environmental degradation through chemical, biological, and photolytic processes.

Because of this, they have been observed to persist in the environment, to be capable of long-range transport, bioaccumulate in human and animal tissue, biomagnify in food chains, and to have potential significant impacts on human health and the environment.

Sources

Sources of air pollution refer to the various locations, activities or factors which are responsible for the releasing of pollutants in the atmosphere. These sources can be classified into two major categories which are:

Anthropogenic sources (human activity) mostly related to burning different kinds of fuel.

- "Stationary Sources" include smoke stacks of power plants, manufacturing facilities (factories) and waste incinerators, as well as furnaces and other types of fuel-burning heating devices.
- "Mobile Sources" include motor vehicles, marine vessels, aircraft and the effect of sound etc.
- Chemicals, dust and controlled burn practices in agriculture and forestry management. Controlled or prescribed burning is a technique sometimes used in forest management, farming, prairie restoration or greenhouse gas abatement. Fire is a natural part of both forest and grassland ecology and controlled fire can be a tool for foresters.

Controlled burning stimulates the germination of some desirable forest trees, thus renewing the forest.

- Fumes from paint, hair spray, varnish, aerosol sprays and other solvents.
- Waste deposition in landfills, which generate methane. Methane is not toxic; however, it is highly flammable and may form explosive mixtures with air. Methane is also an asphyxiant and may displace oxygen in an enclosed space. Asphyxia or suffocation may result if the oxygen concentration is reduced to below 19.5 per cent by displacement
- Military, such as nuclear weapons, toxic gases, germ warfare and rocketry.

Natural Sources

- Dust from natural sources, usually large areas of land with little or no vegetation.
- Methane, emitted by the digestion of food by animals, for example cattle.
- Radon gas from radioactive decay within the Earth's crust. Radon is a colourless, odorless, naturally occurring, radioactive noble gas that is formed from the decay of radium. It is considered to be a health hazard.Radon gas from natural sources can accumulate in buildings, especially in confined areas such as the basement and it is the second most frequent cause of lung cancer, after cigarette smoking.
- Smoke and carbon monoxide from wildfires.
- Volcanic activity, which produce sulfur, chlorine, and ash particulates.

Emission Factors

Air pollutant emission factors are representative values that attempt to relate the quantity of a pollutant released to the ambient air with an activity associated with the release of that pollutant.

These factors are usually expressed as the weight of pollutant divided by a unit weight, volume, distance, or duration of the activity emitting the pollutant (*e.g.*, kilograms of particulate emitted per megagram of coal burned).

Such factors facilitate estimation of emissions from various sources of air pollution. In most cases, these factors are simply averages of all available data of acceptable quality, and are generally assumed to be representative of long-term averages.

The United States Environmental Protection Agency has published a compilation of air pollutant emission factors for a multitude of industrial sources. The United Kingdom, Australia, Canada and other countries have published similar compilations, as has the European Environment Agency.

AIR POLLUTION EFFECTS

We can distinguish between short-term acute effects and long-term chronic effects of air pollution. Acute effects may be brought on, for instance, by pollution episodes, while chronic effects may generally affect a greater proportion of the population. It is important to remember that air pollutants are transboundary, *i.e.* they know no borders and travel easily from their sources towards other locations spreading pollution throughout the world. The respiratory system is by far the most important route for entry of air pollutants into the body and thus becomes their primary target. Air pollution affects virtually everyone but is considered to affect poorer people more than wealthier ones.

AIR POLLUTION EFFECTS ON HUMANS

Air Pollution Effects: Sulfur Dioxide

Sulfur dioxide is a colourless gas with a pungent, suffocating odour. SO_2 is corrosive to organic materials and it irritates the eyes, nose and lungs; therefore it is quite a dangerous air pollutant. Sulfur is contained within all fossil fuels, and is released in the form of sulfur dioxide during fossil fuel combustion.

Because of the widespread use of fossil fuels, sulfur dioxide is among the most common air pollutants produced in every part of the planet. Sulfur dioxide may often act in synergy with other pollutants to produce certain air pollution effects. Let's summarize some sulfur dioxide pollution effects on human health:

- Irritation of eyes, nose, throat; damage to lungs when inhaled
- Acute and chronic asthma
- Bronchitis and emphysema
- Lung cancer

Air Pollution Effects: Nitrogen Dioxide

Nitrogen dioxide is a gas of reddish-brown colour with a distinct sharp, biting odour. It is often analysed in conjunction with another nitrogen gas–nitric oxide. Together these two gases are referred to as NO_x. Combustion of fossil fuels always produces both NO_2 and NO. But almost 90 per cent of the NO_x combustion product is in the form of NO which is then oxidized to NO_2 in the air. NO_x can destroy organic matter, ex. human tissue. Exposure to high concentrations of NO_x can make living organisms more susceptible to bacterial infections and lung cancer. Just like other pollutants, nitrogen dioxide affects people with existing medical conditions more severely than healthy people. For example, asthma sufferers may experience enhanced sensitivity after short-term NO_2 exposure as compared to those without any asthmatic problems. Another group at higher risk is children. For instance, children aged 12 and younger who are exposed to NOx suffer more respiratory illness than the ones who are not exposed.

Depending on different NO_2 concentrations in the air, nitrogen dioxide pollution effects may be as follows:

- Increased incidence of respiratory illness
- Increased airway resistance
- Damage to lung tissue
- Chronic obstructive pulmonary disease, or COPD
- Emphysema
- Pulmonary edema
- Infant and cardiovascular death

It is important to note that nitrogen dioxide is a major component of the photochemical smog and thus a contributor to the formation of ozone which is another serious air pollutant

Air Pollution Effects: Carbon Monoxide

Carbon monoxide is an extremely toxic gas which affects the ability of the body to receive oxygen. Hemoglobin which transports oxygen in the blood is bound by carbon monoxide, which leads to the shortage of oxygen in the body. Carbon monoxide is the most common type of fatal poisoning in many countries around the world. Unlike many other air pollutants, carbon monoxide does not directly affect eyes, nasal passages or lungs.

Exposure to carbon monoxide may lead to:

- Toxicity of the central nervous system and heart
- Headaches, dizziness, nausea and unconsciousness
- Loss of vision
- Decreased muscular coordination
- Abdominal pain
- Severe effects on the baby of a pregnant woman
- Impaired performance on simple psychological tests and arithmetic; loss of judgement of time
- In cases of prolonged exposure to high CO concentrations, unconsciousness, convulsions and death would occur

Carbon monoxide pollution effects are exacerbated in weaker people, specifically in those with heart and lung diseases. What makes this gas even more dangerous is the fact that it is invisible and odourless, and thus not immediately detectable as a hazard.

Air Pollution Effects: Ozone Effects

Ozone is a poisonous gas with a sharp and cold odour. It can be found in the stratosphere where it occurs naturally and plays a beneficial role by protecting the Earth from ultraviolet sunlight; and in the troposphere where it occurs naturally and also forms part of the human-caused photochemical smog. It is of course the tropospheric ozone that we are interested in as an important air pollutant. The chemical reaction between nitrogen oxides and volatile organic compounds in the presence of sunlight results in the photochemical smog; the tropospheric ozone is an end product of this reaction and a component of the smog itself. Because the photochemical smog requires a lot of sunshine to form, it occurs mostly in sunny and heavily polluted places, such as Los Angeles, Mexico City and even Athens. Ozone's main "victim" within the human body is its respiratory system. Once in the lungs, ozone burns through cell walls. The immune system fails to protect the lungs because ozone pushes the defensive cells back. As cellular fluid starts seeping into the lungs, breathing becomes rapid, shallow and painful. Exposure to ozone over long periods of time leads to a stiffening of the lungs and a reduced ability to breathe. As an example, a study conducted in California in the 1980s shows that children living in ozone-polluted areas have smaller than normal lungs and adults lose up to 75 per cent of their lung capacity.

So, exposure to the tropospheric ozone may cause:

- Burning nose and watering eyes
- Tightening of the chest
- Coughing, wheezing and throat irritation
- Rapid, shallow, painful breathing
- Susceptibility to respiratory infections
- Inflammation and damage to the lining of the lungs
- Aggravation of asthma
- Fatigue
- Cancer

Air Pollution Effects: Ammonia Effects

Ammonia is a colourless, pungent, hazardous caustic gas composed of nitrogen and hydrogen. Though ammonia is used for different applications in many sectors, agriculture is its largest consumer and producer. Livestock farming, animal waste and fertilizer application are the biggest sources of atmospheric ammonia emissions within the agricultural sector. Gaseous ammonia is a dangerous air pollutant. Breathing in large amounts can cause death.

Ammonia Pollution Effects

On the respiratory system:

- Nose and throat irritation and burns
- Swelling of the throat and airways; airways destruction
- Pulmonary edema
- Chronic lung disease
- Cough
- Asthma
- Lung fibrosis
- Inhaling large amounts of ammonia can be fatal

On the skin and eyes:

- Skin burns
- Skin conditions, ex. dermatitis
- Burning sensation in the eyes
- Ulceration and perforation of the cornea; blindness
- Cataracts and glaucoma

Air Pollution Effects: Volatile Organic Compounds

Volatile organic compounds are defined as organic compounds which easily evapourate and enter the atmosphere. VOCs are important pollutants for two reasons. First, they are precursors to the formation of ozone; second, they include compounds which are carcinogenic and mutagenic in their own right. Typical VOCs include propane, benzene, ethanol, methanol, ether, carbon tetrachloride and vinyl chloride; substances such as petrol and resins contain many individual VOCs, and many others are produced during combustion processes.

Toxicity of some VOCs and ensuing health effects are no doubt issues of serious concern. For example, exposure to benzene and 1,3-butadiene is a suspected cause of around 10 per cent of leukemia incidence in the UK. Exposure to toluene – another dangerous VOC – may lead to the dysfunction of the central nervous system resulting in behavioural problems, memory loss and disturbance of the circadian rhythm; toluene is also suspected to cause cancer.

Some other VOCs, ex. carbon tetrachloride and PCBs, are believed to produce abnormal changes in fetus development and consequently lead to birth defects. Carbon tetrachloride also leads to liver damage. Vinyl chloride causes Raynaud's phenomenon, necrosis of the small bones of the hand, liver damage, and a rare, highly malignant tumor of the liver.

Benzene pollution effects on human health:

- Short-term breathing of low levels of benzene can cause drowsiness, dizziness, rapid heart rate, headaches, tremors, confusion and unconsciousness; exposure to high levels of benzene may result in death
- *Chronic effects*:
  - — Damage to the bone marrow
  - — Decrease in red blood cells
  - — Excessive bleeding and depression of the immune system increasing the risk of infection
  - — Leukemia

Butadiene pollution effects on human health:

- Short-term exposure to high levels of 1,3-butadiene may lead to distorted blurred vision, vertigo, general tiredness, decreased blood pressure, headache, nausea, decreased pulse rate, and fainting

- Long-term exposure to high levels can increase the risk of cardiovascular diseases and cancer

Let us summarize some effects of VOCs on human health:

- Tiredness, vertigo, drowsiness, dizziness, nausea, confusion, unconsciousness
- Anemia
- Bone marrow damage
- Liver damage
- *Dysfunction of the central nervous system*:
  - — Behavioural problems
  - — Memory loss
  - — Disturbance of the circadian rhythm
- Cardiovascular diseases
- Cancer; specifically leukemia
- Abnormal changes in fetus development, birth defects
- VOCs also contribute to sick building syndrome indoors
- As facilitators in ozone formation, VOCs may indirectly contribute to respiratory problems

AIR POLLUTION EFFECTS: AIRBORNE PARTICLES

Airborne particles are tiny fragments of solid or liquid nature suspended in the air. Particles may be primary – when emitted into the atmosphere directly by sources, or secondary – when formed in the atmosphere through the interaction of primary pollutants. For example, sulfur oxides and nitrogen oxides are primary pollutants which may transform in the air into sulfates and nitrates.

Dust particles are solid particles between 1 and 100 ìm in diametre; fumes, or smoke, are solid particles less than 1 ìm in diametre. Experts believe that dust is the most damaging among all widely measured air pollutants. Smaller dust particles are more dangerous than larger ones because they can penetrate deep into the lungs being deposited on areas where the body's natural clearance mechanisms such as coughing cannot remove them. Combustion of fossil fuels by road transport, power plants etc. is a major source of particulate air pollution. Particles may come in a whole variety of chemical compositions including heavy metals such as cadmium, mercury and lead. They often act in synergy with other air pollutants, ex. sulfur dioxide, to produce negative health effects in humans.

As particulate pollution levels rise, it is common to see an increase in hospital admissions for:

- Stuffy noses, sinusitis
- Sore throats
- Wet cough, dry cough, phlegm
- Head colds
- Burning eyes

- Wheezing; shortness of breath
- Lung disease
- Chest discomfort or pain

As with other pollutants, children are of course more susceptible to particulate pollution.

Specific children's disorders caused by airborne particles may include:

- Infant death
- Low birth weight
- Reduced lung function

Here is a summary of the particulate pollution effects provided by the US Environmental Protection Agency:

- Increased respiratory symptoms
- Decreased lung function
- Aggravated asthma
- Chronic bronchitis
- Irregular heartbeat
- Nonfatal heart attacks
- Premature death in people with heart or lung disease

Indeed, particulate pollution affects both developed and developing countries:

— For example, a major study conducted in the US between 1982 and 1989 found that people living in American cities with the highest levels of particulate air pollution were 15 per cent to 17 per cent more likely to die one to three years early

— In another example, particulate air pollution is linked to nearly 350,000 premature deaths in China and India every year

AIR POLLUTION EFFECTS ON ANIMALS

We don't know a lot about air pollution effects on animals.

- Probably one of the best examples here is that of acid rain and how it affects freshwater animal life.

Sulfur dioxide and nitrogen dioxide are transformed in the atmosphere to produce acid compounds–sulfuric and nitric acids. These compounds then fall back on to the ground as particulates or raindrops—in other words, acid rain. So acid rain also falls on streams and lakes, acidifies them and destroys fish life in these freshwater ecosystems. For example, in Sweden acid rain made over 18,000 lakes so acidic that all the fish died out. Salmon species appear to be particularly sensitive to acidity. Some other populations of animals in Europe and North America that have also been declining due to acid rain are brown trout, mayfly larvae, beetle larvae, mollusks, and aquatic bird species.

- Pollution may also affect animals through plants on which they feed.

For example, pea aphids feed on pea plants exposed to sulfur dioxide in the air. High exposure to sulfur dioxide negatively affects the health of the pea plants, and therefore, the health of the aphids as well.

Some other examples of air pollution effects on animals:

- Excessive ultraviolet radiation coming from the sun through the ozone layer in the upper atmosphere which is eroded by some air pollutants, may cause skin cancer in wildlife
- Tropospheric ozone may damage lung tissues of animals

It is also probably logical to assume that many higher order animals experience air pollution effects similar to those experienced by humans.

AIR POLLUTION EFFECTS ON FORESTS, TREES AND PLANTS

Air pollution can have both long-term and short-term effects on plants.

- Physical injury to leaves is the immediate effect of air pollution on plants. Here is how leaves are affected by different air pollutants:
  - — Ozone produces a speckle of brown spots, which appear on the flat areas of leaf between the veins
  - — *Sulfur dioxide*: Larger bleached-looking areas
  - — *Nitrogen dioxide*: *I*rregular brown or white collapsed lesions on intercostal tissue and near the leaf edge
  - — *Ammonia*: Unnatural green appearance with tissue drying out
- Of all main air pollutants, sulfur dioxide often comes up as the one that most negatively affects plants and trees.

Here is a very illustrative example of how destructive sulfur dioxide can be to vegetation. The development of ore deposits in Canada in the middle of the 20 th century was a major source of sulfur dioxide emissions in high concentrations. Sulfur emissions of one specific iron smelting plant in Wawa, Ontario, caused the destruction of large tracts of native boreal forest in a belt stretching some 30 km from the smelter. Another good example is that of the Black Triangle – the area where the borders of Germany, Poland and the Czech Republic meet, with significant brown coal deposits. Industrial activity in this area led to high sulfur emissions which resulted in devastation of many square kilometres of forest, particularly above an altitude of 750 m. However, it has been noted that sulfur dioxide concentrations in the atmosphere have recently been significantly reduced almost throughout the world and do not carry such dramatic effects as they used to during the 20th century. Lichens are considered to be most sensitive to sulfur dioxide. During the period of high levels of sulfur pollution, large parts of Europe lost many species of lichen and became known as "lichen deserts". But as sulfur pollution levels dropped, many lichen species re-appeared. Sulfur dioxide may also affect higher plants, including wild species, crops and trees.

These effects may be:

- Cell metabolism disruption
- Leaf injury and loss
- Reduced growth and reproduction
- Increase in susceptibility of plants to attacks by insect herbivores

- Nitrogen dioxide, another air pollutant, may act in synergy with sulfur dioxide to produce a negative effect on plants' photosynthesis.
- Tropospheric ozone can prevent plant respiration by blocking stomata and negatively affecting plants' photosynthesis rates which will stunt plant growth; ozone can also decay plant cells directly by entering stomata.
- Particles, just like ozone, often affect plants and trees via blocking of leaf stomata through which plants undertake the gas exchange necessary for photosynthesis and respiration.

Andrew Farmer notes that:

- In sufficient quantities, dusts may form a smothering layer on leaves, reducing light and hence lowering photosynthetic rates. Many dusts are inert and so only act by shading.

However, some dusts are also chemically active. Thus cement dust will also dissolve leaf tissue, resulting in additional injury. Coal dust may also contain toxic compounds. Dusts may also affect ecosystems through their action on soil. Thus the alkaline chemistry of limestone dusts can raise the soil pH of acid and neutral habitats, resulting in the loss of plant and animal species."

So, particulate air pollution effects on plants and trees may be as follows:

- Blocked stomata
- Increased leaf temperature
- Reduced photosynthesis
- Reduced fruit set, leaf growth, pollen growth
- Reduced tree growth
- Leaf necrosis and chlorosis, bark peeling
- Acid rain severely affects trees and plants as well.

It can kill trees, destroy the leaves of plants, can infiltrate soil by making it unsuitable for purposes of nutrition and habitation. It is also associated with the reduction in forest and agricultural yields.

AIR POLLUTION EFFECTS ON WIDER ENVIRONMENT

The effects of acid rain have been known for a long time. Though experts admit that the problem of acid rain is generally under control, it's still worth reiterating the effects brought on by acid rain on the wider environment.

Acid rain has adverse effects on:

- Forests and other vegetation
- Freshwater lakes and streams destroying aquatic life
- Soil
- Buildings and materials

Air pollution also has a negative impact on visibility.

ECONOMIC LOSSES AS AIR POLLUTION EFFECTS

Apart from direct health-related and other environmental issues, air pollution brings with it economic losses as well.

Some of the economic losses caused by air pollution are as follows:

- Direct medical losses
- Lost income from being absent from work
- Decreased productivity
- Travel time losses due to reduced visibility
- Losses from repair of damage to buildings
- Increased costs of cleaning
- Losses due to damage to crops and plants

For example, experts calculated that California is losing $28 billion every year in health care costs, school absences, missed work and lost income potential from premature deaths, because of air pollution in Southern California and in the San Joaquin Valley.

WASTE AND WATER POLLUTION

When toxic substances enter lakes, streams, rivers, oceans, and other water bodies, they get dissolved or lie suspended in water or get deposited on the bed. This results in the pollution of water whereby the quality of the water deteriorates, affecting aquatic ecosystems. Pollutants can also seep down and affect the groundwater deposits. The effects of water pollution are not only devastating to people but also to animals, fish, and birds. Polluted water is unsuitable for drinking, recreation, agriculture, and industry. It diminishes the aesthetic quality of lakes and rivers. More seriously, contaminated water destroys aquatic life and reduces its reproductive ability. Eventually, it is a hazard to human health. Nobody can escape the effects of water pollution.

Importance of Water

Water is our lifeline that bathes us and feeds us. In ancient cultures water represented the very essence of life. The Romans were the first to pipe water into their growing cities, especially with their aqueducts. They also realised that sewage water could cause damage to their people, and needed to be removed from large areas of people. Water has played a role not only in the history of countries, but in religion, mythology, and art.

Water in many religions cleanses the soul through holy water.The water at Lourdes, France is thought by many religions to be sacred water with healing powers. In Egyptian mythology, the Nu was the beginning of everything and represented water. It brought life to their people, but in drought, produced chaos. The water or hydrologic cycle explains interactions between the atmosphere, hydrosphere, and lithosphere. The water or hydrologic cycle is a major driving force on our planet. Water is in constant motion, evapourating into the atmosphere from oceans, lakes, rivers and streams.

When the atmosphere can no longer support the moisture within the clouds, we experience rain, snow, hail, or sleet. Some water is locked in the form of ice at the polar caps and in glaciers. Water melts in the spring, producing run-off, that percolates through the Earth as groundwater (subsurface) or makes its way back to the sea (surface). The oceans contain most of the water, but it is salt water which is unusable by most organisms. Only pure H_2O (water) can interact with organisms. The movement of the oceans also has a direct effect on the atmosphere.

The atmosphere is that envelope of gas that keeps organisms living on this planet. Oceans and atmosphere interact to give us weather. Water provides the Earth with the capacity of supporting life. An organism doesn't have to be told how important water is to their existence. An amphibian knows to lay their eggs in water or else there will be no new born. Even flies know to lay their eggs in fresh water.The only organism that doesn't understand the importance of water is humans, especially in industrialized countries.

Children in those societies turn on the water in a sink and never think about the trouble someone has gone for that "miracle" to occur. Dams, reservoirs, filtering plants, and pipes all bring clean water when the facet is turned on. Sewage water is only mixed with recycled water supplies after the water goes through rigorous cleaning methods. Water borne diseases are any illnesses caused by drinking contaminated water. Diseases can include infection from bacteria (Salmonella), viruses, or by small parasites (Cryptosporida, Giardia, and Toxoplasma). These organisms and viruses cause diseases like cholera, typhoid fever, malaria, botulism, polio, dysentery, giardia, and hepatitis. A one of the first symptoms of these diseases is diarrhea, which cause about three million deaths throughout the world.Sewage is sometimes discharged into rivers, where children downstream might be taking a bath or using the water to drink. The simplest treatment method is boiling. Just bring the water to a boil for at least one minute, then allow it to cool. But this is not always effective in heavily chemical polluted water supplies. Without water, organisms could not exist.

Water is a resource that should not be taken for granted. It needs to be conserved, just as we save other valuable resources. Water is one of the weirdest compounds known to humans. The difference between the boiling point and freezing point of water is one of the largest ranges of any compound. It is this span of temperature that mirrors the range of where life can exist, from bacteria to humans. Water also has a very high specific heat, which means that it can absorb or lose much heat before its temperature changes. This is important in maintaining body heat in mammals.

It also takes a lot of energy before vapourization can occur. For this reason, water evapourates slowly from ponds and lakes, where many life forms are dependent on a stable, warm environment. Water is less dense in its solid state than in its liquid state, so that ice floats instead of sinking. This property permits life to develop in polar and subpolar regions where ice floats and allows life to continue living below the surface. If ice were heavier than water, it would sink, and more ice would form on top of it. As a result, all life in the waters would be trapped in the ice in the many areas of the world where it gets cold enough to freeze water.

Resistance of Water to a Disturbance

Water also exhibits viscosity. One can observe the effects of viscosity alongside a stream or river with uniform banks. The water along the banks is nearly still, while the current in the centre may be swift. This resistance between the layers is called viscosity. This property allows smaller fish to live near the shore, while larger fish are able to swim efficiently in strong currents. Viscosity is also responsible for the formation of eddies, creating turbulence that leads to good mixing of air in the water and more uniform distribution of microscopic organisms. Domestic sewage refers to waste water that is discarded from households.

Also referred to as sanitary sewage, such water contains a wide variety of dissolved and suspended impurities. It amounts to a very small fraction of the sewage by weight. But it is large by volume and contains impurities such as organic materials and plant nutrients that tend to rot. The main organic materials are food and vegetable waste, plant nutrient come from chemical soaps, washing powders, etc. Domestic sewage is also very likely to contain disease-causing microbes. Thus, disposal of domestic waste water is a significant technical problem.

Industrial Effluents

Waste water from manufacturing or chemical processes in industries contributes to water pollution. Industrial waste water usually contains specific and readily identifiable chemical compounds. During the last fifty years, the number of industries in India has grown rapidly. But water pollution is concentrated within a few subsectors, mainly in the form of toxic wastes and organic pollutants. Out of this a large portion can be traced to the processing of industrial chemicals and to the food products industry. In fact, a amount of large- and medium-sized industries in the region covered by the Ganga Action Plan do not have adequate effluent treatment facilities. Most of these defaulting industries are sugar mills, distilleries, leather processing industries, and thermal power stations. Most major industries have treatment facilities for industrial effluents. But this is not the case with small-scale industries, which cannot afford enormous investments in pollution control equipment as their profit margin is very slender.

Effects of Water Pollution

The effects of water pollution are not only devastating to people but also to animals, fish, and birds. Polluted water is unsuitable for drinking, recreation, agriculture, and industry. It diminishes the aesthetic quality of lakes and rivers. More seriously, contaminated water destroys aquatic life and reduces its reproductive ability. Eventually, it is a hazard to human health. Nobody can escape the effects of water pollution.

Effects of Run-off Pollution

Rain picks up dirt and silt and carries it into the water. If the dirt and silt settle in the water body, then these sediments prevent sunlight from reaching aquatic plants. If the Sun can't reach the plants, these perish. These sediments also clog fish gills and smother organisms that live on the bottom of the body of the water.

Effects of Oil Pollution and Antifreeze

If oil is spilled on the water, the effects on the ecosystem and the components are harmful. Many animals can be annihilated in case they ingest oil. Oil contaminated prey may be a reason of death for many. If the oil coats the feathers of birds, these may die. Oil and antifreeze makes the water have a foul odour and there is a sticky film on the surface of water that kills animals. Oil is the most harmful pollutant in the water.

Contaminated Ground Water Effects

If contaminated water enters the ground, there may be serious effects. People may become very sick and there is a probability of developing liver or kidney problems and cancer or other illnesses.

Fertilizers and Other Chemicals

Nitrates in drinking water leads to diseases of infants that may lead to their death. Cadmium is a metal in sludge-derived fertilizer. This can be absorbed by crops. When people ingest this, they may cause diarrheal disorders, liver and kidney damage. The inorganic substances like mercury, arsenic and lead are the causes of pollution. Other chemicals can also lead to problems concerning the taste, smell and colour of water. Pesticides, PCBs and PCPs are all poisonous to all sorts of life. Pesticides are used in farming, homes and forestry. PCBs are found as insulators in old electrical transformers. PCPs are found in products like wood preservatives.

Effects of Agricultural Water Pollution

Rain and irrigation water drains off cultivated land that has been fertilized and treated with pesticides, the excess nitrogen and poisons are mixed with it into the water supply. These pesticides are toxic and pollute the water in a different mode. Aquatic plants growth cause deoxyge-nation of water and annihilate flora and fauna in a stream, lake and river. Fertilizers enhance the growth of bacteria that are in water and increase the concentration of bacteria to hazardous levels.

Effects of Thermal Water Pollution

Machinery in the industries are cooled with water from lakes and rivers. This water reaches the river in a heated state. This water decreases the ability of the aquatic system to hold oxygen and raises the growth of warm water species.

Effects of Heavy Metal Water Pollution

Heavy metals like lead, mercury, iron, cadmium, aluminum and magnesium are present in water sources. If these metals are present in the sediment, these reach the food chain through plants and aquatic animals. This causes heavy metal poisoning in case the level in the water is very high.

BIOREACTORS FOR WASTE WATER TREATMENT

A bioreactor may refer to any device or system that supports a biologically active environment. In one case, a bioreactor is a vessel in which is carried out a chemical process which involves organisms or biochemically active substances derived from such organisms. This process can either be aerobic or anaerobic. These bioreactors are commonly cylindrical, ranging in size from some litre to cube meters, and are often made of stainless steel.

A bioreactor may also refer to a device or system meant to grow cells or tissues in the context of cell culture. These devices are being developed for use in tissue engineering. On the basis of mode of operation, a bioreactor may be classified as batch, fed batch or continuous. An example of a bioreactor is the chemostat. Organism growing in bioreactor may be suspended or immobilized. The simplest, where cells are immobilized, is a Petri dish with agar gegel. Large scale immobilized cell bioreactors are:

- Packed bed
- Fibrous bed
- Membrane

Bioreactor Design

Bioreactor design is quite a complex engineering task. Under optimum conditions the microorganisms or cells are able to perform their desired function with great efficiency. The bioreactor's environmental conditions like gas (*i.e.*, air, oxygen, nitrogen, carbon dioxide) flowrates, temperature, pH and dissolved oxygen levels, and agitation speed/circulation rate need to be closely monitored and controlled.

Most industrial bioreactor manufacturers use vessels, sensors, controllers, and a control system, networked together for their bioreactor system, see programmable logic controller (PLC).

Fouling can harm the overall sterility and efficiency of the bioreactor, especially the heat exchangers. To avoid it the bioreactor must be easily cleanable and must be as smooth as possible (therefore the round shape).

A heat exchanger is needed to maintain the bioprocess at a constant temperature. Biological fermentation is a major source of heat, therefore in most cases bioreactors need water refrigeration. They can be refrigerated with an external jacket or, for very large vessels, with internal coils.

In an aerobic process, optimal oxygen transfer is perhaps the most difficult task to accomplish. Oxygen is poorly soluble in water -and even less in fermentation broths- and is relatively scarce in air (20.8 per cent). Oxygen transfer is usually helped by agitation, that is also needed to mix nutrients and to keep the fermentation homogeneous. There are however limits to the speed of agitation, due both to high power consumption (which is proportional to the cube of the speed of the electric motor) and the damage to organisms due to excessive tip speed.

Industrial bioreactors usually employ bacteria or other simple organisms that can withstand the forces of agitation. They are also simple to sustain, requiring only simple nutrient solutions and can grow at astounding rates.

In bioreactors where the goal is grow cells or tissues for experimental or therapeutic purposes, the design is significantly different from industrial bioreactors. Many cells and tissues, especially mammalian, must have a surface or other structural support in order to grow, and agitated environments are often destructive to these cell types and tissues. Higher organisms also need more complex growth medium.

NASA has developed a new type of bioreactor that artificially grows tissue in cell cultures. NASA's tissue bioreactor can grow heart tissue, skeletal tissue, ligaments, cancer tissue for study, and other types of tissue.

BIOLOGICAL TREATMENT FOR WASTEWATER

Industrial wastewater treatment covers the mechanisms and processes used to treat waters that have been contaminated in some way by man's industrial or commercial activities prior to its release into the environment or its re-use.

Most industries produce some wet waste although recent trends in the developed world have been to minimise such production or recycle such waste within the production process. However, many industries remain dependent on processes that produce a water based waste stream.

It is a processes, which employ aerobic or anaerobic microorganisms and result in decanted effluents and separated sludge containing microbial mass together with pollutants. Biological treatment processes are also used in combination and/or in conjunction with mechanical and advanced unit operations.

Increasing water pollution and the pathetic state of rivers worldwide demands a wastewater plant that can handle aerobic degradation of waste. Eco-friendliness means, in quintessence, effective pollution control both in the air and on the ground – yet while belching smokestacks are now rarely seen, pollution control on the ground is still a serious cause for concern for the industry, and as a result effluent or wastewater treatment has become mandatory for all industries.

Industrial wastewater treatment is a group of unit processes designed to separate, modify, remove, and destroy undesirable substances carried by wastewater from industrial sources. United States governmental regulations have been issued that involve volatile organic substances, designated priority pollutants; aquatic toxicity as defined by a bioassay; and in some cases nitrogen and phosphorus. As a result, sophisticated technology and process controls have been developed for industrial wastewater treatment.

Wastewater streams that are toxic or refractory should be treated at the source, and there are a number of technologies available. For example, wet air oxidation of organic materials at high temperature and pressure (2000 lb/in. or 14 kilopascals and 550°F or 288°C) is restricted to very high concentrations of these substances. Macroreticular (macroporous) resins are specific for the removal of particular organic materials, and the resin is regenerated and used again. Membrane processes, particularly reverse osmosis, are high-pressure operations in which water passes through a semipermeable membrane, leaving the contaminants in a concentrate.

Pretreatment and primary treatment processes address the problems of equalization, neutralization, removal of oil and grease, removal of suspended solids, and precipitation of heavy metals. Aerobic biological treatment is employed for the removal of biodegradable organics. An aerated lagoon system is applicable (where large land areas are available) for treating non-toxic wastewaters, such as generated by pulp and paper mills. Fixed-film processes include the trickling filter and the rotating biological contactor. In these processes, a biofilm is generated on a surface, usually plastic. As the wastewater passes over the film, organics diffuse into the film, where they are biodegraded. Anaerobic processes are sometimes employed before aerobic processes for the treatment of high-strength, readily degradable wastewaters. The primary advantages of the anaerobic process is low sludge production and the generation of energy in the form of methane (CH_4) gas.

Biological processes can remove only degradable organics. Non-degradable organics can be present in the influent wastewater or be generated as oxidation by-products in the biological process. Many of these organics are toxic to aquatic life and must be removed from the effluent before discharge. The most common technology to achieve this objective is adsorption on activated carbon.

In some cases, toxic and refractory organics can be pretreated by chemical oxidation using ozone, catalyzed hydrogen peroxide, or advanced oxidation processes. In this case the objective is not mineralization of the organics but detoxification and enhanced biodegradability.

Biological nitrogen removal, both nitrification and denitrification, is employed for removal of ammonia from wastewaters. While this process is predictable in the case of municipal wastewaters, many industrial wastewaters are inhibitory to the nitrifying organisms.

Volatile organics can be removed by air or steam stripping. Air stripping is achieved by using packed or tray towers in which air and water counterflow through the tower. In steam stripping, the liquid effluent from the column is separated as an azeotropic mixture. Virtually all of the processes employed for industrial wastewater treatment generate a sludge that requires some means of disposal. In general, the processes employed for thickening and dewatering are the same as those used in municipal wastewater treatment. Waste activated sludge is usually stabilized by aerobic digestion in which the degradable solids are oxidized by prolonged aeration.

Most landfill leachates have high and variable concentrations of organic and inorganic substances. All municipal and most industrial landfill leachates are amenable to biological treatment and can be treated anaerobically or aerobically, depending on the effluent quality desired. Activated carbon has been employed to remove non-degradable organics. In Europe, some plants employ reverse osmosis to produce a high-quality effluent.

AEROBIC TREATMENT SYSTEM

An aerobic treatment system or ATS, often called (incorrectly) an aerobic septic sytem is a small scale sewage treatment system similar to a septic tank system, but which uses an aerobic process for digestion rather than the anaerobic process used in septic systems. These systems are commonly found in rural areas where public sewers are not available, and may be used for a single residence or for a small group of homes.

Process

The ATS process generally consists of the following phases:

- Pre-treatment stage to remove large solids and other undesireable substances from the wastewater.
- Aeration stage, where the aerobic bacteria digest the biological wastes in the wastewater.

- Settling stage to allow any undigested solids to settle. This forms a sludge which must be periocally removed from the system.
- Disinfecting stage, where chlorine or similar disinfectant is mixed with the water, to produce an antiseptic output.

Comparison to Tradional Septic Systems

The aeration state and the disinfecting stage are the primary differences from a traditional septic system to the user of the system. These stages increase the initial cost of the aerobic system, and also the maintainence requirements over the passive septic system. The aerator requires a supply of electricity to drive the air pump, and the disinfectant must be periodically renewed. On the positive side, the aerobic system produces a much higher quality effluent, and the leach field can be far smaller than that of a similar capacity septic system, and the output can be discharged in areas too environmentally sensitive for septic system output. Some aerobic systems recycle the effluent through a sprinkler system, using it to water the lawn.

Aerobic systems are similar to conventional septic systems in that they both use natural processes to treat wastewater. But unlike septic anaerobic treatment, the aerobic treatment process requires oxygen. Aerobic treatment units, therefore, use a mechanism to inject and circulate air inside the treatment tank. This mechanism requires electricity to operate. For this reason, aerobic systems cost more to operate and need more routine maintenance than most septic systems. However, when properly operated and maintained, aerobic systems can provide a high quality wastewater treatment alternative to septic systems.

Why Use Aerobic Treatment

Aerobic Treatment Units (ATUs) have proven to be a reliable technology when they are properly designed, constructed, and maintained. When properly monitored and maintained their performance is consistent. ATUs may enable development or use of difficult sites. They can remedy existing malfunctioning systems and they can be a good option for homes in environmentally sensitive areas.

Process Description

Aerobic systems treat wastewater using natural processes that require oxygen. Bacteria that thrive in oxygen-rich environments work to break down and digest the wastewater inside the aerobic treatment unit.

Like most onsite systems, aerobic systems treat the wastewater in stages. Sometimes the wastewater receives pretreatment before it enters the aerobic unit, and the treated wastewater leaving the unit requires additional treatment or disinfection before being returned to the environment. Such a variety of designs exists for home aerobic units and systems that it is impossible to describe a typical system. Instead, it is more practical to discuss how some common design features of aerobic systems work and the different stages of aerobic treatment.

Pretreatment

Some aerobic systems include a pretreatment step to reduce the amount of solids in the wastewater going into the aerobic unit. Solids include greases, oils, toilet paper, and other materials that are put down the drain or flushed into the system. Too much solid material can clog the unit and prevent effective treatment.

Some pretreatment methods include a septic tank, a primary settling compartment in the pretreatment unit, or a trash trap. Pretreatment is optional but can greatly improve a unit's performance.

Aerobic Treatment Units

The main function of the aerobic treatment unit is to collect and treat household wastewater, which includes all water from toilets, bathtubs, showers, sinks, and laundry. Aerobic units themselves come in many sizes and shapes-rectangular, conical, and some shapes that defy classification. There are two typical aerobic treatment designs:

1. Suspended growth units and
2. Attached growth units.

The process most aerobic units use to treat wastewater is referred to as suspended growth. These units include a main compartment called an aeration chamber in which air is mixed with the wastewater. Since most home aerobic units are buried underground like septic tanks, the air must be forced into the aeration chamber by an air blower or through liquid agitation.

The forced air mixes with wastewater in the aeration chamber, and the oxygen supports the growth of aerobic bacterial that digests the solids in the wastewater. This mixture of wastewater and oxygen is called the mixed liquor.

The treatment occurring in the mixed liquor is referred to as suspended growth because the bacterial grow as they are suspended in the liquid unattached to any surface.

Unfortunately, the bacterial cannot digest all of the solids in the mixed liquor, and these solids eventually settle out as sludge. Many aerobic units include a secondary chamber called a settling chamber or clarifier where excess solids can settle. Other designs allow the sludge to accumulate at the bottom of the tank.

In aerobic units designed with a separate settling compartment, the sludge returns to the aeration chamber (either by gravity or by a pumping device). The sludge contains bacterial that also aid in the treatment process. Although, in theory, the aerobic treatment process should eventually be able to consume the sludge completely, in practice, the sludge does build up in most units and will need to be pumped out periodically so that solids don't clog the unit.

An alternative design for aerobic treatment is the attached growth system. These units treat wastewater by taking a surface made of material that the bacteria can attach to, and then exposing that surface alternately to wastewater and air. This is done either by rotating the surface in and out of the wastewater or by dosing the wastewater onto the surface. Pretreatment is required. The air needed for the process either is naturally present or is supplied mechanically.

Attached growth systems, such as trickling filters and rotating disks, are less common than suspended growth systems, but have certain advantages. For example, there is no need for mixing, and solids are less likely to be washed out of the system during periods of heavy household water use.

Flow Design

The way and the rate in which wastewater is received by and flows through the aerobic unit differs from design to design. Continuous flow designs simply allow the wastewater to flow through the unit at the same rate that it leaves the home. Other designs employ devices (such as pretreatment tanks, surge chambers, and baffles) to control the amount of the incoming flow. Batch process designs use pumps or siphons to control the amount of wastewater in the aeration tank and/or to discharge the treated wastewater in controlled amounts after a certain period.

Controlling the flow of wastewater helps to protect the treatment process. When too much wastewater is flushed into the system all at once, it can become overburdened, and the quality of treatment can suffer. The disadvantages to mechanical flow control devices are that, like all mechanical components, they need maintenance and run the risk of malfunctioning.

Disinfection

Some units have the disinfection process incorporated into the unit design. In some cases, disinfection may be the only treatment required of the wastewater from an aerobic unit before the water is released into the environment. Added costs for disinfectants, such as chlorine, should be taken into account with aerobic units.

Design Criteria

Aerobic units should be large enough to allow enough time for the solids to settle and for the wastewater to be treated. The daily wastewater volume is usually determined by the number of bedrooms in the house. Flows in Arizona for individual homes are up to 140 gallons per day (gpd) per bedroom.

The needed size of an aerobic unit is often estimated the same way the size of a septic tank is estimated, by the number of bedrooms (not bathrooms) in the house. It is assumed that each person will use approximately 50 to 100 gallons of water per day, and that each bedroom can accommodate two people.

When calculated this way, a three-bedroom house will require a unit with a capacity of 300 to 600 gallons per day.

Some health departments require that aerobic units be sized at least as large as a septic tank in case the aerobic unit malfunctions and oxygen doesn't mix with the wastewater. In such cases, the aerobic unit will work as a septic tank, which will, at least, provide partial treatment for the wastewater.

Lower temperatures tend to slow down most biological processes, and higher temperatures tend to speed them up. The aerobic process itself creates heat, which, along with the heat from the electrical components, may help to keep the treatment process active. However, cold weather can have adverse effects on the performance of aerobic units.

In one 1977 study of aerobic units, bulking of the sludge seemed to occur when the temperature of the mixed liquor fell below 15 degrees Celsius (59 degrees Fahrenheit). Problems can sometimes be avoided by insulating around the units.

Operation and Maintenance

Aerobic treatment systems are not accepted in all areas. Regulations for onsite systems can vary from state to state and county to county. A major reason that aerobic systems are not more widely used is concern about improper operation and maintenance by homeowners. Aerobic systems require regular maintenance, and abuse or neglect can easily lead to system failure.

A typical inspection should include removing the unit's cover and checking its general appearance. Check pipes and the inside of the aeration chamber and note the appearance of the wastewater inside the unit and its colour and odour. If the unit includes a chlorinator, this too will need to be checked and may need cleaning. Samples may be taken of the mixed liquor from the aeration chamber, as well as the final treated wastewater. Check to see that all mechanical parts, alarms, and controls are in working order, and that solids are pumped from the system if needed.

It is important that mechanical components in aerobic systems receive regular inspection and maintenance. For example, air compressors sometimes need to be oiled, and vanes, filters, and seals may need to be replaced. Malfunctions are common during the first few months after installation. In most cases, homeowners do not have the expertise to inspect, repair, and maintain their own systems.

Most aerobic units have controls that can be switched on and off by the homeowner in case of emergency. Aerobic units also are required to have alarms to alert the homeowner of malfunctions. Depending on the design of the system, controls and alarms can be located either inside or outside the home, and alarms can be visible, audible, or both. Homeowners should make sure that controls and alarms are always protected from corrosion, and that the aerobic unit is turned back on if here is a power outage or if it is turned off temporarily.

To assure homeowners are receiving a reputable ATU most states require the system to by approved by the NSF International (formerly the National Sanitation Foundation). NSF has tested aerobic units according to the requirements of ANSI/NSF Standard 40. NSF is a non-profit organization devoted to the protection of the environment through the development of product standards, product evaluations, research, education and training. The American National Standards Institute (ANSI) is the recognized accredited in the U.S. for organizations that develop consumer standards and for those that provide independent product evaluations. NSF is accredited by ANSI for both of these areas of service.

ANAEROBIC DIGESTION

Anaerobic digestion (AD) is the harnessed and contained, naturally occurring process of anaerobic decomposition. An anaerobic digester is an industrial system that harnesses these natural process to treat waste, produce biogas that can be used to power electricity generators, provide heat and produce soil improving material.

Anaerobic digesters have been around for a long time and they are commonly used for sewage treatment or for managing animal waste. Increasing environmental pressures on waste disposal has increased the use of AD as a process for reducing waste volumes and generating useful byproducts. It is a fairly simple process that can greatly reduce the amount of organic matter which might otherwise end up in landfills or waste incinerators.

Almost any organic material can be processed in this manner. This includes biodegradable waste materials such as waste paper, grass clippings, leftover food, sewage and animal waste. Anaerobic digesters can also be fed with specially grown energy crops to boost biodegradable content and hence increase biogas production. After sorting or screening to remove inorganic or hazardous materials such as metals and plastics, the material to be processed is often shredded, minced, or hydrocrushed to increase the surface area available to microbes in the digesters and hence increase the speed of digestion. The material is then fed into an airtight digester often with extra water added depending on the digestion process and feedstock.

MICROBIAL FUEL CELL

A microbial fuel cell (MFC) or biological fuel cell is a device in which a chemical, typically glucose, is converted to electric power by means of bacteria on the anode side.

Power outputs are usually small, in the order of magnitude of about a milliwatt, and there are no current commercially available applications.

However, some hope to use them in the future to build a glucose-powered pacemakers that would need no other power supply than the glucose present in the bloodstream. A microbial Fuel Cell is a device that converts chemical energy to electrical energy by the catalytic reaction of microorganisms. A typical microbial fuel cell consists of anode and cathode compartments separated by a cation specific membrane. In the anode compartment, fuel is oxidized by microorganisms, generating electrons and protons. Electrons are transferred to the cathode compartment through, and the protons through the membrane. Electrons and protons are consumed in the cathode compartment reducing oxygen to water. In general, there are two types of microbial fuel cell, mediator- and mediator-less microbial fuel cell. Biological fuel cells take glucose and methanol from food scraps and convert it into hydrogen and food for the bacteria.

Mediator Microbial Fuel Cell

Most of the microbial cells are electrochemically inactive. The electron transfer from microbial cells to the electrode is facilitated by mediators such as potassium ferric cyanide, thionine, methyl viologen (methyl blue), humic acid, neutral red and so on (Delaney et al., 1984; Lithgow et al., 1986). Most of the mediators available are expensive and toxic.

Mediator-less Microbial Fuel Cell

Mediator-less microbial fuel cells have been engineered at the Korea Institute of Science and Technology, by a team led by Kim, Byung Hong A mediator-less microbial fuel cell does not require a mediator but uses electrochemically active bacteria to transfer electrons to the electrode (electrons are carried directly from the bacterial respiratory enzyme to the electrode). Among the electrochemically active bacteria are, *Shewanella putrefaciens* (Kim *et al.*, 1999a), *Aeromonas hydrophila* (Cuong *et al.*, 2003), and others.

Mediator-less MFCs are a much more recent development and due to this the factors that affect optimum operation, such as the bacteria used in the system, the type of ion membrane, and the system conditions such as temperature, are not particularly well understood. Bacteria in mediator-less MFCs typically have electrochemically-active redox enzymes such as cytochromes on their outer membrane that can transfer electrons to external materials.

6

Risk Assessment of Nanomaterials

ADVANCES IN NANOMATERIALS

The history of nanomaterials began immediately after the big bang when Nanostructures were formed in the early meteorites. Nature later evolved many other Nanostructures like seashells, skeletons etc. Nanoscaled smoke particles were formed during the use of fire by early humans. The scientific story of nanomaterials however began much later. One of the first scientific report is the colloidal gold particles synthesised by Michael Faraday as early as 1857. Nanostructured catalysts have also been investigated for over 70 years. By the early 1940's, precipitated and fumed silica nanoparticles were being manufactured and sold in USA and Germany as substitutes for ultrafine carbon black for rubber reinforcements.

Nanosized amorphous silica particles have found large-scale applications in many every-day consumer products, ranging from non-diary coffee creamer to automobile tires, optical fibres and catalyst supports. In the 1960s and 1970's metallic nanopowders for magnetic recording tapes were developed. In 1976, for the first time, nanocrystals produced by the now popular inert-gas evapouration technique was published by Granqvist and Buhrman. Recently it has been found that the Maya blue paint is a nanostructured hybrid material.

The origin of its colour and its resistance to acids and biocorrosion are still not understood but studies of authentic samples from Jaina Island show that the material is made of needle-shaped palygorskite crystals that form a superlattice with a period of 1.4 nm, with intercalates of amorphous silicate substrate containing inclusions of metal nanoparticles. The beautiful tone of the blue colour is obtained only when both these nanoparticles and the superlattice are present, as has been shown by the fabrication of synthetic samples.

Today nanophase engineering expands in a rapidly growing number of structural and functional materials, both inorganic and organic, allowing to manipulate mechanical, catalytic, electric, magnetic, optical and electronic functions. The production of nanophase or cluster-assembled materials is usually based upon the creation of separated small clusters which then are fused into a bulk-like material or on their embedding into compact liquid or solid matrix materials. *e.g.* nanophase silicon, which differs from normal silicon in physical and electronic properties, could be applied to macroscopic semiconductor processes to create new devices. For instance, when ordinary glass is doped with quantized semiconductor "colloids," it becomes a high performance optical medium with potential applications in optical computing.

METHODS FOR CREATING NANOSTRUCTURES

There are many different ways of creating nanostructures: of course, macromolecules or nanoparticles or buckyballs or nanotubes and so on can be synthesized artificially for certain specific materials.

They can also be arranged by methods based on equilibrium or near-equilibrium thermodynamics such as methods of self-organization and self-assembly.

Using these methods, synthesized materials can be arranged into useful shapes so that finally the material can be applied to a certain application.

MECHANICAL GRINDING

Mechanical attrition is a typical example of 'top down' method of synthesis of nanomaterials, where the material is prepared not by cluster assembly but by the structural decomposition of coarser-grained structures as the result of severe plastic deformation.

This has become a popular method to make nanocrystalline materials because of its simplicity, the relatively inexpensive equipment needed, and the applicability to essentially the synthesis of all classes of materials. The major advantage often quoted is the possibility for easily scaling up to tonnage quantities of material for various applications.

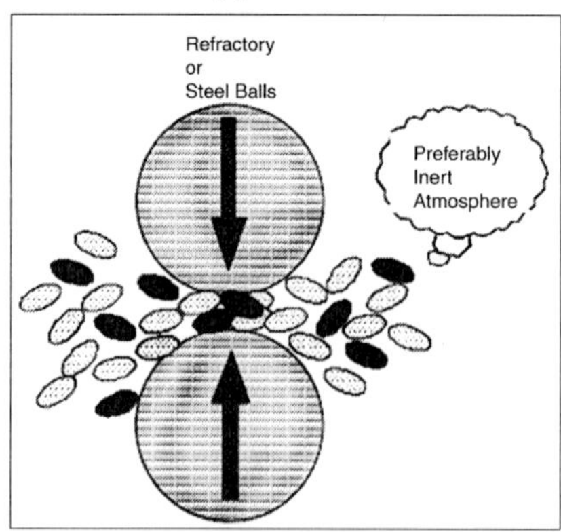

Fig. Schematic Representation of the Principle of Mechanical Milling

Similarly, the serious problems that are usually cited are:

- Contamination from milling media and/or atmosphere, and
- To consolidate the powder product without coarsening the nanocrystalline microstructure.

In fact, the contamination problem is often given as a reason to dismiss the method, at least for some materials. Here as suggested, review the mechanisms presently believed responsible for formation of nanocrystalline structures by mechanical attrition of single phase powders, mechanical alloying of dissimilar powders, and mechanical crystallisation of amorphous materials. The two important problems of contamination and powder consolidation will be briefly considered.

Mechanical milling is typically achieved using high energy shaker, planetary ball, or tumbler mills. The energy transferred to the powder from refractory or steel balls depends on the rotational speed, size and number of the balls, ratio of the ball to powder mass, the time of milling and the milling atmosphere. Nanoparticles are produced by the shear action during grinding. Milling in cryogenic liquids can greatly increase the brittleness of the powders influencing the fracture process. As with any process that produces fine particles, an adequate step to prevent oxidation is necessary.

Hence this process is very restrictive for the production of non-oxide materials since then it requires that the milling take place in an inert atmosphere and that the powder particles be handled in an appropriate vacuum system or glove box. This method of synthesis is suitable for producing amorphous or nanocrystalline alloy particles, elemental or compound powders. If the mechanical milling imparts sufficient energy to the constituent powders a homogeneous alloy can be formed. Based on the energy of the milling process and thermodynamic properties of the constituents the alloy can be rendered amorphous by this processing.

Wet Chemical Synthesis of Nanomaterials

In principle we can classify the wet chemical synthesis of nanomaterials into two broad groups:

- The top down method: where single crystals are etched in an aqueous solution for producing nanomaterials, For example, the synthesis of porous silicon by electrochemical etching.
- The bottom up method: consisting of sol-gel method, precipitation etc. where materials containing the desired precursors are mixed in a controlled fashion to form a colloidal solution.

Sol-gel Process

The sol-gel process, involves the evolution of inorganic networks through the formation of a colloidal suspension and gelation of the sol to form a network in a continuous liquid phase. The precursors for synthesizing these colloids consist usually of a metal or metalloid element surrounded by various reactive ligands. The starting material is processed to form a dispersible oxide and forms a sol in contact with water or dilute acid. Removal of the liquid from the sol yields the gel, and the sol/gel transition controls the particle size and shape. Calcination of the gel produces the oxide. Sol-gel processing refers to the hydrolysis and condensation of alkoxide-based precursors such as $Si(OEt)_4$. The reactions involved in the sol-gel chemistry based on the hydrolysis and condensation of metal alkoxides M(OR)z can be described as follows:

$$MOR + H_2O \qquad MOH + ROH \text{ (Hydrolysis)}$$

$$MOH + ROM \qquad M\text{-}O\text{-}M + ROH \text{ (Condensation)}$$

Sol-gel method of synthesizing nanomaterials is very popular amongst chemists and is widely employed to prepare oxide materials. The sol-gel process can be characterized by a series of distinct steps.

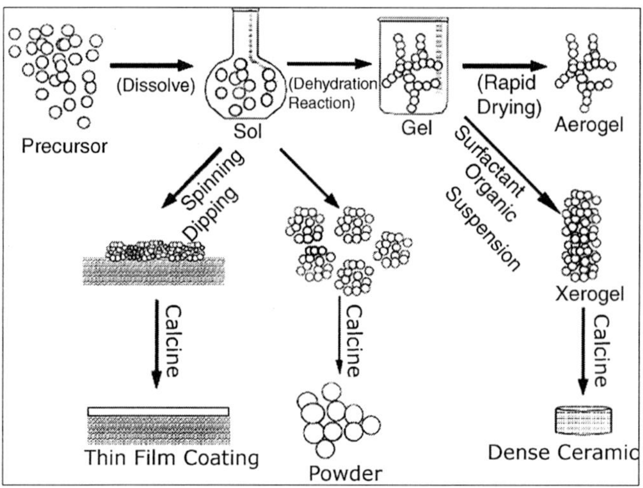

Fig. Schematic Representation of Sol-gel Process of Synthesis of Nanomaterials

- Formation of different stable solutions of the alkoxide or solvated metal precursor.
- Gelation resulting from the formation of an oxide-or alcohol-bridged network by a polycondensation reaction that results in a dramatic increase in the viscocity of the solution.
- Aging of the gel, during which the polycondensation reactions continue until the gel transforms into a solid mass, accompanied by contraction of the gel network and expulsion of solvent from gel pores. Ostwald ripening and phase transformations may occur concurrently with syneresis. The aging process of gels can exceed 7 days and is critical to the prevention of cracks in gels that have been cast.

- Drying of the gel, when water and other volatile liquids are removed from the gel network. This process is complicated due to fundamental changes in the structure of the gel. The drying process has itself been broken into four distinct steps:
  - The constant rate period,
  - The critical point,
  - The falling rate period, the second falling rate period.

  If isolated by thermal evaporation, the resulting monolith is termed a *xerogel*. If the solvent (such as water) is extracted under supercritical or near super critical conditions, the product is an *aerogel*.
- Dehydration, during which surface-bound M-OH groups are removed, there by stabilizing the gel against rehydration. This is normally achieved by calcining the monolith at temperatures up to 800°C.
- Densification and decomposition of the gels at high temperatures (T>800°C). The pores of the gel network are collapsed, and remaining organic species are volatilized.

The interest in this synthesis method arises due to the possibility of synthesizing non-metallic inorganic materials like glasses, glass ceramics or ceramic materials at very low temperatures compared to the high temperature process required by melting glass or firing ceramics.

The major difficulties to overcome in developing a successful bottom-up approach is controlling the growth of the particles and then stopping the newly formed particles from agglomerating. Other technical issues are ensuring the reactions are complete so that no unwanted reactant is left on the product and completely removing any growth aids that may have been used in the process. Also production rates of nano powders are very low by this process. The main advantage is one can get monosized nano particles by any bottom up approach.

Gas Phase Synthesis of Nanomaterials

The gas-phase synthesis methods are of increasing interest because they allow elegant way to control process parameters in order to be able to produce size, shape and chemical composition controlled nanostructures. Before we discuss a few selected pathways for gas-phase formation of nanomaterials, some general aspects of gas-phase synthesis needs to be discussed.

In conventional chemical vapour deposition (CVD) synthesis, gaseous products either are allowed to react homogeneously or heterogeneously depending on a particular application:

- In homogeneous CVD, particles form in the gas phase and diffuse towards a cold surface due to thermophoretic forces, and can either be scrapped of from the cold surface to give nano-powders, or deposited onto a substrate to yield what is called *'particulate films'*.

- In heterogeneous CVD, the solid is formed on the substrate surface, which catalyses the reaction and a dense film is formed.

In order to form nanomaterials several modified CVD methods have been developed.

Gas phase processes have inherent advantages, some of which are noted here:

- An excellent control of size, shape, crystallinity and chemical composition
- Highly pure materials can be obtained
- Multicomonent systems are relatively easy to form
- Easy control of the reaction mechanisms

Most of the synthesis routes are based on the production of small clusters that can aggregate to form nano particles. Condensation occurs only when the vapour is supersaturated and in these processes homogeneous nucleation in the gas phase is utilised to form particles. This can be achieved both by physical and chemical methods.

Furnace

The simplest fashion to produce nanoparticles is by heating the desired material in a heatresistant crucible containing the desired material. This method is appropriate only for materials that have a high vapour pressure at the heated temperatures that can be as high as 2000°C. Energy is normally introduced into the precursor by arc heating, electronbeam heating or Joule heating. The atoms are evaporated into an atmosphere, which is either inert or reactive. To carry out reactive synthesis, materials with very low vapour pressure have to be fed into the furnace in the form of a suitable precursor such as organometallics, which decompose in the furnace to produce a condensable material. The hot atoms of the evaporated matter lose energy by collision with the atoms of the cold gas and undergo condensation into small clusters via homogeneous nucleation.

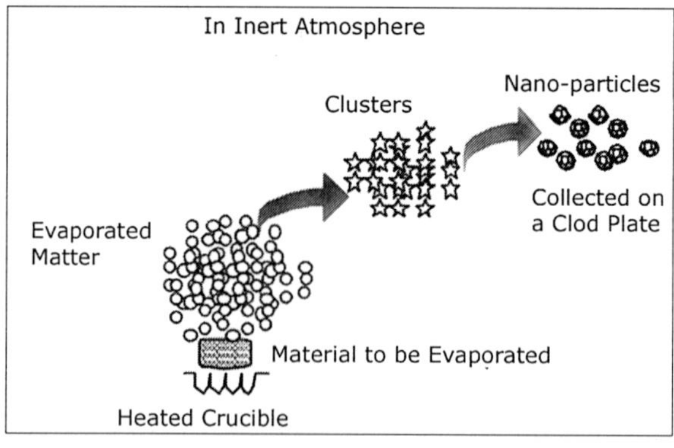

Fig. Schematic Representation of Gas Phase Process of Synthesis of Single Phase Nanomaterials from a Heated Crucible

In case a compound is being synthesized, these precursors react in the gas phase and form a compound with the material that is separately injected in the reaction chamber. The clusters would continue to grow if they remain in the supersaturated region. To control their size, they need to be rapidly removed from the supersaturated environment by a carrier gas.

The cluster size and its distribution are controlled by only three parameters:

- The rate of evaporation (energy input),
- The rate of condensation (energy removal), and
- The rate of gas flow (cluster removal).

Because of its inherent simplicity, it is possible to scale up this process from laboratory (mg/day) to industrial scales (tons/day).

IMPACT OF NANOMATERIALS

Although nanotechnology will help to reduce environmental and human health hazards, for example through better solar cells, replacement of hazardous chemicals, improved pollutant extraction or prevention technologies, unfavourable safety aspects and concerns need also be addressed. However, this is a complex task, as differently shaped nanoparticles of the same substance may have very different properties.

Looking beyond the potential technical risks associated with nanomaterials, there is actually only scarce information available about the impact of nanomaterials on non-human species, on ecosystems or the global environment. Even within established questions of toxicology we still do not know how different nanoparticles exactly interact in the human body or the environment, which means that more data is needed.

There is a particular need to address the potential impact of nanomaterials along the lifecycle of a product, from manufacture through disposal ("cradle to grave"). And this 'life cycle view' is not only necessary regarding possible ecotoxicological effects in the different stages of the life cycle of the product, but also regarding the (dissipative) use or consumption of energy and materials (which is in the focus of life cycle assessment (LCA)).

Another problem arises from the fact that the existing relevant regulatory framework (chemicals, wastes, safety) may be adequate for some specific areas, where only small amounts of nanomaterials are used (like in research laboratories), but may not be adequate for the industrial mass production and use of these materials. European legislation1 obliges employers to take measures necessary to ensure the safety and health protection of workers, which should also apply in the case of handling nanomaterials. There are specific Directives relating to the risks to workers due to exposure to carcinogens and mutagens2, chemical agents3, the use of workplace equipment and the use of personal protective equipment5. Under the new REACH chemical regulations, manufacturers have a duty to ensure that substances placed on the market do not adversely affect human health and the environment.

While there is no explicit mention of nanomaterials, they would in principle be covered by these regulations. The Integrated Pollution Prevention and Control Directive requires industrial installations to limit emissions and would be applicable also to nanomaterials. The Waste Directive requires Member States to take measures to ensure that waste treatment does not have an adverse impact on health and the environment.

Consequently, there is a need to evaluate the extent to which existing regulatory and associated risk assessment frameworks (strategies, methodologies and tools), can be applied, or extended to cover mass production, application, and final disposal of nanomaterials and associated products. Looking at REACH there must be a solution regarding the fact, that some important regulations in REACH are triggered by mass (in tons).

This is inadequate for nanomaterials, because for some engineered nanoparticles their number and/or shape could be of higher relevance. Additionally there is a demand for adaptation of the OECD-Guidelines (for hazardous (chemical) substances) to the testing of nanomaterials. But we cannot wait until all these data will be collected. We need processes for 'preliminary assessment' that deliver 'preliminary information as the base of precautionary measures.

The latter may be based on information about chemical and or physical properties, on quantitative structure activity relations (QSAR), on the probability of exposure (*e.g.* because of extreme mobility and or persistence), and measures based on it may reach from precautionary measures, aiming at exposure avoidance or reduction, up to principles for a safer (or sustainable) design of nanomaterials and products. Along with the growing (eco)toxicological knowledge about impacts, this 'preliminary knowledge' will become less important.

When working with nanomaterials, there are four main aspects of their life-cycle that must be considered: material selection, manufacturing, application, and disposal/recycle. Although most work focuses on the possible toxic effects of nanocomponents after exposure for risk assessment, it is worth examining the potential contribution of these materials to all impacts listed above when they are added to products or processes, to better understand the importance of the underlying choices that are involved with the implementation of a nanotechnology.

Therefore, understanding the toxicity of nanomaterials and nano-enabled products is important for human and environmental health and safety as well as public acceptance. In numerous toxicological studies it was shown that nanoparticles have implications on the human health inducing, e. g., pulmonary and systemic inflammation. It was also shown that nanoparticles inhaled can translocate to different parts within the human body, including the brain. Data available on the (eco)toxicity of nanomaterials is still limited, but studies prove that there are toxic effects on wildlife and a potential for bioaccumulation in various organisms.

DESTINY OF NANOMATERIALS

Our current knowledge and the available scientific data on nanoparticles characterisation, detection and measurement, their transport, toxicology, exposure and persistence is still insufficient, to allow an accurate and reliable assessment of their final fate in the environment.

But the increasing amount of nanomaterials produced world-wide raises issues on their destiny when released into the environment and on possible hazards due to accumulation in animals, plants and the human body. Metallic nanoparticles may be extremely resistant to degradation and may accumulate in waters or soils. They may aggregate, which in turn will change their properties compared to single nanoparticles, be transported and accumulate in soils, groundwater and sediments. Existing regulations are based on parametres that may not be appropriate for nanoparticles in solution or in suspension.

Concentration data alone is inadequate to quantify the true exposure to nanomaterials, but needs accurate measurement of other nano-specific parametres, like surface area and reactivity. More appropriate analytical methods are required to reliably detect nanoparticles in various compartments and measure their physico-chemical properties in air, water, soil, and consumer products, the media in which humans and ecosystems are exposed to. Better analytical techniques are required to detect these particles in cells, fluids, or plant tissues.

Existing methods to assess environmental exposure levels are not appropriate to determine their environmental fate and current risk assessment procedures need to be modified for adequate hazard characterization. There is a distinct need of data on the properties, toxicokinetics and degradability of nanoparticles to better understand where, in which form and to what extent nanoparticles will end up in the environment, to develop more accurate impact assessment models and to find efficient solutions for product design that favours their reuse and recyclability.

The proposed NanoSustain project will address this need for more reliable data on the fate of nanoparticles by developing technical solutions and advanced computer-based models to support the assessment of the hazard, distribution and fate of nanomaterials that may be released from products.

OCCUPATIONAL HAZARDS OF NANOMATERIALS

The environmental, health and safety concerns of nanomaterials in fact span from research safety, manufacturing safety, to consumer (product) safety, and include environmental impacts throughout the life cycle, until final disposal of the nanomaterial products and wastes. Among these concerns, research and pilot scale production safety is most relevant to the research and development work at HKUST. Due to the rapid development and many uncharted areas in nanoscience and nanotechnology, the practice of occupational health and safety also faces many challenges. As for many conventional workplace contaminants, airborne route is considered the most crucial for worker protection.

The shape and size of airborne nanoparticles, and the conglomeration status are some of the main factors determining the hazard. Nanoparticles in liquid phase may present skin hazard when contacted.

The hazard of the liquid matrix, its potential interaction with the nanoparticles, *e.g.* serving as a carrier that facilitates skin penetration, and the conglomeration status of the particles will affect the overall hazard. In either the airborne or liquid phase, the nanomatierals have the potential to adversely impact the environment if they are not properly contained during usage and disposed of afterwards.

CHALLENGES IN ASSESSING AND CONTROLLING

A major challenge in evaluating hazard posed by airborne nanoparticles is to find a (set of) suitable exposure metric(s). It will be necessary to consider number concentration and surface area, other than the conventional mass concentration, then appropriate sampling and analytical methods, and eventually health-based standards, will have to be developed. Quite a few number concentration direct reading instruments originally designed for ultrafine particle monitoring are available, but their applicability to monitoring of nanoparticles in a work environment remains to be tested. Also the reliability of the detection technology and how "direct" the reading really is are also main concerns to be addressed. In order to differentiate engineered nanoparticles of interest with natural or incidental nanosized particles that may be present in the background, examination by transmission electron microscopy (TEM) remains the golden standard for characterizing and enumerating nanoparticles.

At least one passive sampling method coupled with enumeration by electron microscope has been developed, its precision and accuracy are yet to be determined. Governmental agencies of developed countries such as the National Institute for Occupational Safety and Health (NIOSH) of the US, and Health and Safety Executive (HSE) of the UK, are actively working to fill up these knowledge gaps in assessing hazards of airborne nanoparticles.

Another major challenge is control technologies. Traditionally, high efficiency particulate air (HEPA) filters are used for removing airborne particulates. However, before the rise of nanomaterial research, it has never been proved whether HEPA filter can capture nano scale particles. Theoretically the filtering efficiency of HEPA increases with decreasing size below 0.1-0.3 μm, which is the most difficult size range to capture according to physics principles of filtration by fibrous materials. Previously there were only measured data points down to 0.01 μm (10 nm), beyond which the filtration efficiency was by extrapolation.

However, according to preliminary data presented in the Second International Symposium on Nanotechnology and Occupational Health (Minneapolis, October 2005) by a group of researchers from University of Minnesota, 3M and NIOSH, HEPA filter is capable of removing more than 99.97 per cent of particles down to the range of 2-4 nm. No significant thermal rebound was observed. This means with our current knowledge, HEPA filter can be deployed as an effective means to contain and control airborne nanoparticles at least down to the low nm size range.

As for skin protection, most of the glove and protective clothing materials currently in use have not been tested against nanosize particles. Presumably the materials should be protective against nanoparticle conglomerates. Some suggested treating airborne nanoparticles as toxic gas in terms of personnel protection. In any case, until information on interaction of nanoparticles and protective material is available, skin protection should at least be adequate to prevent penetration of the carrier medium of nanoparticles in the liquid phase.

Either in air or liquid phase, nanoparticles tend to agglomerate instead of existing in single particle form. This is helpful from a hazard control point of view. However, certain future applications or manufacturing processes may rely on dispersed nanoparticles. Therefore it is prudent to plan safety measures assuming the worst case scenario.

Some basic precautionary measures have been suggested by organizations such as HSE of UK and NIOSH of US. These include treating all nanomaterials that may possess unique physical chemical properties as "new chemicals" for regulatory purpose.

Material and safety information raw material can only be viewed, at best, as a starting point or reference for assessing the risk of the nanomaterial product. A good example is carbon, which is practically non-toxic as a parent material; but a group of scientists have already established a toxicity level of oral lethal dose 50 (LD50) of 300 mg/kg for C_{60} buckyballs, which is considered moderately toxic.

An important concept that worth mentioning is "Responsible Nanotechnology", which basically says, there are many promising applications for nanoscience and nanotechnology, however, there are also uncertainties in potential health and environmental effects.

Therefore we should not repeat the errors we made before in cases like DDT and chlorofluorocarbons, where people focused on the beneficial use of the new chemicals so much that they did not carefully check for potential health and environmental impacts until damages were done. Concurrent development of technology and environment, health and safety assessment and controls must go hand in hand in this potentially revolutionary field of research.

The UK Royal Society and Royal Academy of Engineering published a report titled "Nanoscience and nanotechnologies: opportunities and uncertainties" in July 2004.

They proposed the following general guidelines, which form a good foundation for responsible nanotechnology development:

- Treat nanomaterials as if they are hazardous
- Seek to remove them from waste stream
- Don't apply free nanomaterials to environment unless benefit clearly outweigh risk
- Assess potential impact throughout lifecycle of nanomaterial

In early 2006, the Government supported the establishment of Nanotechnology and Advanced Material Incorporated (NAMI) under the direction of HKUST to facilitate nanoresearch in Hong Kong and transfer of nanotechnology to local industries. This undoubtedly elevates HKUST to a leading position in nanotechnology development in Hong Kong. It also brings into focus the need to integrate EHS assessment into nanotechnology development all the way from research, pilot production, manufacturing, consumer use and final disposal. This presents a great opportunity for us to realise the idea of "Responsible Nanotechnology", *i.e.* to not only take advantage of the benefits that nanotechnology brings us, but also ensure that we do not enjoy these benefits in the expense of health, safety or the environment.

HKUST

Nanomaterial and nanotechnology is a rapidly evolving field, and is one of the focus areas of research at HKUST. The environmental health and safety knowledge about nanomaterials and nanotechnology is also being generated by many researchers worldwide. Even though there are considerable uncertainties, it is prudent to take precautionary measures to minimize the potential of adverse health, safety, and environmental impacts. The following are elements of the Nanosafety Programme at HKUST. Due to rapid development of this field, these programme elements will be revised when necessary.

- *Facility Review*: HSEO has been involved in the planning of laboratory facility for the Institute of Nanomaterials and Nanotechnology (INMT). When faculty members establish other research facilities for nanomaterials and nanotechnology, they should involve HSEO at the planning stage to ensure necessary safety installations are included in the physical facility, safe operational protocols are developed, and suitable personal protective equipment is prescribed and made available to lab workers.
- *Research Proposal Review*: A nanosafety section has been added to the research proposal safety review form under the Committee of Research Practices (CRP), where principal investigators are requested

to declare whether any nanomaterial will be used, and if so, the physical and chemical properties of the nanomaterial, known toxicity, use conditions, proposed control measures, and final disposal arrangement. Based on the information provided, the Safety Panel of CRP will suggest necessary precautionary measures.

- *Laboratory Visit*: At this early stage of development, one of the crucial elements is to establish a database of nanoresearch activities on campus. Based on the information collected from the research proposal review, HSEO staff will visit research groups working on nanoscience and nanotechnology to collect information on the nanomaterials, work processes, potential hazards involved, and waste disposal practices. Since safety standard is yet to be developed, available guidelines such as those from NIOSH and HSE will form the basis of recommendations on EHS issues.
- *Safety Precautions and Safety Installations*: As mentioned in Section E above, HEPA filter is a proven method to contain and control airborne nanoparticles. Research should conduct all processes that may give rise to airborne nanoparticles inside one of the following safety devices:
  - Biological safety cabinet (BSC), if a dust-free environment is not critical, a Class 1 BSC can be used. Otherwise, a Class 2 BSC can provide both a dust-free environment and protect lab workers at the same time. However, BSC cannot contain volatile or toxic chemicals, if these are to be used together with nanoparticles. One should use a fume hood equipped with HEPA filtration
  - Fume hood equipped with HEPA filtration
  - Glove box

  Be reminded that BSC, fume hood and HEPA filter all need periodic testing to ensure they are properly functioning. Please contact either FMO-LS or HSEO if you have questions about such periodic testing.
- Safety Precautions and Personal Protective Equipment (PPE)
  Laboratory workers handling nanoparticles or other nanomaterials should follow these guidelines:
- Always ensure personal hygiene and observe good general laboratory practices, including no drinking and eating in the laboratory, always wearing lab coat and safety glasses, etc.
- Consider the hazards of the parent materials in assessing hazards of nanomaterials. However, remember the nano features of the materials most likely will present many other physicochemical and toxicological properties not found in the parent materials.
- Never handle nanomaterials with bare hands or expose such materials to bare skin. Until further information is available, use chemical

resistant gloves and protective clothing. The material of these PPE items should at least be protective against the carrier medium, such as the solvent, containing the nanomaterials.

- If it is necessary to handle nanoparticles outside of a safety installations specified in the previous section, use appropriate PPE such as a properly fitted elastomeric respirator with HEPA filter cartridge. Please contact HSEO for requirements of a respirator user. If the exposure risk to nanoparticles is considered low, particulate respirators with a filtration efficiency equivalent to HEPA (*e.g.* the 100 series approved by NIOSH or P3 by EC standard) may be used.
- A vacuum cleaner equipped with a HEPA filter may be used to remove nanomaterials from reactors or work areas. However, the integrity of the HEPA filter must be periodically tested to ensure proper functioning.
- Dispersed airborne nanoparticles, like many other airborne fine particles, can become a fire and explosion hazard. Care must be taken to prevent ignition.
- Equipment in contact with nanomaterials should be evaluated for residual hazard before servicing, reuse for other purposes, or disposal.
- Contact HSEO for assistance in evaluating or controlling hazards related to nanomaterials.
- Waste Disposal
  - Waste containing nanomaterials should be disposed as chemical wastes as a minimum. See Chapter 16 for general chemical waste disposal procedures.
  - Researchers should seek to remove the nanomaterials from the waste stream where possible.
  - Alternatively, researchers should destruct the nano features of the materials where feasible. For examples, acid may be added to waste solution containing quantum dots to dissolve the nanoparticles before disposal.
  - Following the principle of "Responsible Nanotechnology, and the concept of minimizing potential environmental impact, researchers should develop suitable destruction method for nanomaterials as a part of their research.
  - Contact HSEO for assistance in disposal of nanomaterial waste.
- *Exposure Monitoring*: Even though standard method has not been established for measuring exposure to nanoparticles yet, some conventional exposure monitoring methods can be employed to provide an estimate of exposure. Some field instruments can also be used to measure airborne nanoparticles. HSEO will continue to monitor the development of exposure measuring technology for nanomaterials.

- *Continuous Improvement*: As the field is rapidly evolving, HSEO will continue to monitor the development, especially the environment, health and safety aspects, and will introduce necessary amendments to the nanosafety programme. We urge all researchers to integrate EHS considerations into their research projects, and contribute their expertise to the responsible development of nanotechnology at HKUST.

The continuing revolution in computer hardware is packing ever more logic gates in ever smaller volumes. Unfortunately this produces ever more heat, which will limit the feasible packing density and performance unless the energy dissipated by each logic operation can also be dramatically reduced.

To reduce the energy dissipation of a logic operation below ln(2) kT (near thermal noise) requires the use of reversible logic for fundamental thermodynamic reasons. Extrapolation of current trends suggest this limit will become significant within one to two decades. Many real devices can be viewed as electrically controlled switches, so a method of using an abstract switch in a reversible manner is useful. Two methods of using switches to implement reversible computations are discussed in this paper.

The first method has an energy dissipation which is proportional to the square of the error in the voltage, while the second method has an energy dissipation which can in principle be reduced indefinitely by slowing the speed of computation.

The first method is basically an extension to "pass logic" which has been previously used with both nMOS (hot clock nMOS) and CMOS transmission gates to achieve low energy dissipation. The second method is a novel thermodynamically reversible logic system based on CCD-like operations which switches charge packets in a reversible fashion to achieve low energy dissipation. There is now a fairly extensive literature on reversible computation which shows that the energy dissipation per device operation cannot be reduced below ln(2) kT (where k is Boltzman's constant and T is the temperature; kT is roughly the thermal energy of a single atom) if the device is not reversible.

For the last 50 years the energy dissipation per gate operation has been declining with remarkable regularity. Extrapolation of this trend shows the energy dissipation per device operation reaching kT by the year 2015. (This assumes that T is 300 Kelvin-more on this later).

To gain some perspective on this consider that an "AND" gate which has a power supply of one volt and which allows 100 electrons to go from that one volt supply to ground during the course of a switching operation will dissipate 100 electron volts.

Although 100 electron volts is about 4,000 times kT (and well above the theoretical limit), it will be difficult for simple improvements of irreversible approaches to reach even this level of energy dissipation.

Extrapolating present trends, we should reach 4,000 kT before the year 2000, *e.g.*, within ten years. While reversible logic is an absolute necessity if we are to reduce energy dissipation per device operation below kT, it is a useful heuristic for designing systems that have low energy dissipation even when the actual energy dissipation is well above kT.

Even if we do develop irreversible devices that approach ln(2) kT, a computer operating at room temperature at a frequency of one gigahertz with 10 ^ 18 logic gates packed into a volume not much larger than a cubic centimeter would dissipate over three megawatts. The drive for ever greater computational power with ever more densely packed logic elements will eventually require that a single logic operation dissipate orders of magnitude energy less than kT.

If we are to realise the full potential of nanoelectronics and molecular logic devices, at some point in the future we will be forced to use thermodynamically reversible logic elements for a substantial fraction of logic operations.

We can state quite confidently that one of three things will occur:

- The historic rate of decreasing energy dissipation per device operation will slow or halt in the next one or two decades or
- We will operate computers at lower temperatures or
- We will develop new methods of computing in a reversible way that can beat the kT barrier.

The first option is unattractive. First, cooling is a major problem. The heroic cooling methods used in the Cray 3 supercomputer to remove the heat generated by the computer's operation suggest that failure to reduce energy dissipation per gate operation would be a major limiting factor in future computer performance.

While both air-cooling a single chip that dissipates 150 watts and water cooling of a chip dissipating 790 watts/cm ^ 2 have been demonstrated, and more effective cooling methods should be feasible in the future, we will eventually reach a limit to heat removal.

Second, in many applications power is limited. Portable computers, remote sensors and other isolated systems have only a limited amount of power available. Third, although the raw cost of electrical power is not yet a major limitation it would become so in the future if reductions in energy dissipation did not keep pace with advances in other areas.

The Wall Street Journal said "Computer systems currently account for 5 per cent of commercial electricity use in the U.S., with the potential to grow to 10 per cent by the year 2000." Operating computers at a lower temperature will not reduce overall energy dissipation. If we operated future devices at 3 Kelvins we would reduce kT by a factor of 100 and so could reduce energy dissipation per logic operation by a similar factor-but for fundamental thermodynamic reasons the coefficient of performance of the refrigerator for the system can be at best 3K/(300K–3K) or about 0.01.

Thus, the lower energy required per gate operation will be almost exactly balanced by the increased energy needed by the refrigerator. In many applications refrigeration is not an attractive option. Lap top and portable computers, embedded systems, various "smart" appliances and other applications make the use of refrigeration undesirable. Factors other than net energy savings can make low temperature operation worthwhile. For example, some potentially attractive devices don't operate at higher temperatures.

In large computers operating in stable environments (the traditional computer centre, for example) refrigeration might be attractive, particularly if it permits the use of devices that provide much better performance but which require low temperature for their operation (Josephson junction devices might be an example). Refrigeration *per se*, however, does not seem too attractive. Finally, and most attractively, we could use reversible computation to reduce energy dissipation. This would, in principle, allow energy dissipations indefinitely below kT per logic operation.

While some barrier should eventually be encountered, the use of reversible logic should allow us to continue current trends in energy dissipation per logic operation for the longest possible time. It is interesting to note that certain uses of pass logic, transmission gates, and hot clock nMOS do in fact perform some logic operations in a thermodynamically reversible fashion.

Research in the design and utilization of reversible logic to futher lower energy dissipation will simply recognize and make explicit a trend that has already begun. Further research in this area is the most appropriate response to the rather limited range of possibilities that face us.

REVERSIBLE ARCHITECTURES

One concern about this approach is the need to use reversible computer architectures. Such architectures are entirely feasible. A wide variety of computations can be done in a logically reversible manner. Bennett concluded: "...for any e > 0, ordinary multitape Turing machines using time T and space S can be simulated by reversible ones using time O(T ^ (1+e)) and space O(S log T) or in linear time and space O(S T ^ e)." This general result shows that even arbitrary irreversible computations can be mapped into reversible computations.

Specific irreversible computations can often be mapped into reversible computations with little or no loss of efficiency. Even if we do not adopt new reversible computer architectures, simple applications of reversible computation could be made within the framework of existing architectures.

A typical computer executes a sequence of instructions, and each instruction will typically change the contents of a single register or memory location. Although loading the result of the instruction into a register will normally be irreversible (it erases the previous contents of the register) it is still the case that all other operations performed by the computer during instruction execution could in principle be made reversible.

Thus, although we would have to dissipate roughly kT energy for each bit in the output register for each instruction execution, we need not use dissipative logic throughout the computer (as is currently done).We can go one step further without making significant changes in computer architecture: the simple register-register add instruction $R_1 = R_1 + R_2$ is logically reversible. With proper hardware, this particular instruction could in principle be made to dissipate as little energy as we desired. While non-reversible instructions (*e.g.*, R1 = 0) would still dissipate greater energy, the overall energy dissipation of the computer could be reduced significantly if a reversible approach were developed and used to implement the reversible instructions. Of course, once the energy-wasting instructions were identified, compilers would learn to avoid using them. Today, compilers can optimize for speed of computation and for the amount of memory used.

In the future, compilers could also optimize a programme to minimize the energy dissipated during computation, *e.g.*, to minimize the number of bits that are erased. This would provide an entirely evolutionary path from the current irreversible designs to computer architectures that were as reversible as was practically feasible. While it is perhaps not obvious how far this trend can go it is clear that a very large percentage of computer operations can be made reversible-perhaps a remarkably large percentage.

REVERSIBLE DEVICES

Several reversible devices have been proposed over the years. Von Neumann and independently Goto proposed the parametron which encodes information in the phase of an oscillation. Fredkin and Toffoli proposed an electronic reversible logic family which used switches, capacitors and inductors. By turning the switches on and off at just the right times, charge could be transferred between the capacitors through the inductors with minimal energy loss.

By increasing the size of the inductors it would be possible to create a family of ever slower but ever more energy efficient circuits. As the LC time constant is increased, the energy dissipation can be decreased to an arbitrary extent.

A proposal by Likharev based on Josephson junctions "...is particularly significant, because it is a genuine example... of a system that has frictional forces proportional to velocity" according to Landauer. The parametric quantron is based on Josephson junctions and operates at low temperatures. It should dissipate less than kT when operating at 4 kelvins and with a switching time of 1 nanosecond. Modern "high temperature" superconductors might allow operation at relatively high temperatures, including that of liquid nitrogen.

Even these temperatures would still be a significant disadvantage. Molecular-scale mechanical reversible "rod logic" has been proposed by Drexler which should dissipate less than kT when operating at room temperature at a speed of 50 to 100 nanoseconds. Drexler's proposal is also significantly smaller than the parametric quantron. While the ability to fabricate Drexler's proposal is some decades away, it provides a good argument that molecular-scale reversible devices should be feasible.

Although the parametric quantron can be used in a reversible fashion to achieve low energy dissipation it uses magnetic fields to couple logic devices, requires low temperatures to operate and will likely prove resistant to scaling to the molecular size range. By contrast, Drexler's mechanical proposal might well prove smaller than even the smallest of future electronic devices. Mechanical devices depend fundamentally on the position of the nuclei of the atoms of which they are made. The position of the nucleus of an atom can be much more precisely known than the position of an electron: quantum uncertainty in the position of the electron is much greater because the mass of the electron is much smaller.

If device function depends on the position of an electron then device size will be larger than if device function depends on the position of the nucleus. Such molecular mechanical proposals will, however, almost certainly be slower because of the greater mass of the nuclei of atoms as compared with the mass of electrons. In this paper we discuss simple methods of using relatively conventional electronic switching devices in a reversible fashion which can be scaled to a very small size. The methods described here allow a simple voltage-controlled switch to be used in a reversible manner, both to implement combinational logic and also to iterate an arbitrary reversible logic function.

Other methods are feasible. For example, helical logic—an elegant form of reversible electronic logic—has recently been proposed by Merkle and Drexler. Voltage controlled switches of various types have long been used in electronic devices. Not all voltage controlled switches are suitable for use in low energy designs because of their intrinsic energy dissipation.

Traditional relays actuated by electromagnets dissipate comparatively large amounts of energy. Even if a switch were suited for low energy operation it would still dissipate energy if it were turned on when there was a voltage across it. The abrupt rush of current that normally occurs when a switch is turned on would insure this. Most previous logic designs based on switches operated in this manner (though the proposal by Fredkin and Toffoli avoids this problem, and the proposal by Seitz et. al. eliminates dissipative switch closures for combinational logic, although not for sequential logic).

The switch is turned on for the purpose of draining charge when the charge is no longer wanted. That is, a logic "1" might be encoded by a charge on a capacitor. Turning on a switch to drain away that charge would dissipate energy even if the switch itself were perfectly non-dissipative. A related but distinct line of research also emphasizes the importance of energy dissipation, but assumes that the losses involved in dissipatively switching a wire from one logic state to another are fixed and cannot be altered, *e.g.*, that slowing the switching speed will not reduce energy dissipation.

Even assuming that this is true, clever design of the circuit can greatly reduce the number of times it is necessary to take such a dissipative action. Further research which combines both this approach and the approaches described here would seem fruitful. Even relatively recently some authors have argued that switching devices must fundamentally dissipate more than kT regardless of how slowly they operate. It is worth noting, therefore, that charging and discharging a condenser need not dissipate the energy involved.

For example, a simple LC oscillator will repeatedly charge and discharge a condenser. Although the energy stored on the condenser is 1/2 CV ^ 2, this energy need not be dissipated during a charge/discharge cycle.

If the Q of the circuit is high the energy dissipation per cycle will be small. In addition, the energy dissipation of such an oscillator depends on frequency. By using larger condensers and inductors the frequency of operation can be slowed and the fraction of energy dissipated per cycle reduced. If all linear dimensions of a device are scaled in size in proportion to some characteristic length L then capacitance and inductance are proportional to L while frequency and resistance are proportional to 1/L. The Q of the circuit is proportional to L so the percentage of the total energy in the oscillator that is dissipated per cycle can be reduced simply by increasing all linear dimensions equally.

Different scaling methods can produce better results. If we assume that a fixed capacitive load is charged through a fixed resistance by a voltage that is ramped up over a period of time T, then the energy dissipated by the resistance is approximately 1/2 CV ^ 2 × 2RC/T (for RC << T).

This dissipation can be reduced as much as we might want by increasing T (and leaving the circuit components unchanged). Koller and Athas have called this adiabatic charging. We will call it "thermodynamically reversible" or simply "reversible," relying on context to make it clear that we are referring to an asymptotically non-dissipative process. Note that the term "reversible" can also be used to mean "logically reversible" with no implication that it is asymptotically non-dissipative.

A logically reversible process might be highly dissipative, so it is sometimes necessary to distinguish between "logically reversible" and "thermodynamically reversible. The situation is quite different for a step function in which the voltage is ramped up instantly (T=0).

This process dissipates 1/2 CV ^ 2. Charge falls through the fixed potential drop in a dissipative fashion much as water goes over Niagra falls. Although this is the method in common use today to charge and discharge the various capacitive loads in computer circuits, it is not the method that we will be considering.

By using reversible charging, further increases in the switching time can be used to decrease the energy dissipation per logic operation. This, of course, results in a slower speed of operation leading to a trade-off between speed and energy dissipation. How to arrange matters so that current flows smoothly and reversibly is the principle topic of the rest of the paper.

This is challenging because the basic switching devices are:

- Abrupt and discontinuous and
- Pass through a dissipative intermediate state (which is neither fully on nor fully off) while switching.

THE FIRST METHOD

The first method uses a conventional voltage controlled switch but in a somewhat unusual manner. In essence we never turn a switch on or off when there is either a voltage across it or current going through it. This mode of operation is reversible. We extend this approach to implement a sequential circuit which can iterate the computation of a reversible logic function. It is also related to "dry switching," a method for using relays which avoids turning relays on or off when there is a voltage across them. The primary purpose of dry switching is to avoid deterioration of the relay contacts, rather than minimize switching energy.

We also adopt the following convention for turning a switch on and off: when the control input to the switch is at logic "0" (low) the switch is turned on and current can flow from the input to the output or from the output to the input. When the control input is "1" (high) the switch is turned off and current cannot flow. The reader should note that this convention is not the usual one. We adopt it instead of its opposite —where a "1" turns the switch on and a "0" turns it off— because it is more intuitive when applied to the second proposal discussed later in this paper, the CCD-based reversible logic element.

The proposed mode of operation is fundamentally clocked and multi-staged. The outputs of stage i drive the inputs of stage i+1. For the moment, we assume the output of stage 0 is simply given. That is, a set of input lines are assumed to have fixed voltages upon them representing the input data.

The value of the succesive logic levels can then be computed in turn. The logic values (voltages) produced by stage 1 are computed from the logic values given by stage 0; The logic values produced by stage 2 are computed from the logic values produced by stage 1; etc. Simple pass logic can be employed to implement combinational logic. A "NOR" operation in this logic is simply two switches in series. A clock signal (initially at logic 0) is connected to the input of the first switch.

If the control inputs of both switches have settled to logic 0, then both switches will conduct. The clock input connected to the first switch is slowly raised (from logic 0 to logic 1). Both switches will conduct and the output will gradually shift to a logic 1. If either input is 1, one of the switches will not conduct and the output will remain at 0.

MAKING AN N-LEVEL CIRCUIT

We assume that we have an N level combinational circuit, and that N "clock signals" are provided. Initially, clocks 1 through N are low (logic 0), and the outputs of stage 0 are at some preset values. The outputs from stage 1 are computed as simple logical functions from the outputs of stage 0. The outputs from stage i are used as the inputs to stage i+1. A "NOR" and "NAND" gate from stage 1. The NAND gate is logically complete so any arbitrary boolean function can be implemented by using an appropriately connected collection of them. The clock signals are now raised in turn.

The output of "stage 0" is assumed to be valid prior to the start of operations, and clock 0 is unspecified (for now). Clock 1 is raised, and the outputs from stage 1 become valid. Clock 2 is raised, and the outputs from stage 2 become valid. This sequence continues, until clock N is raised, and the outputs from stage N become valid. A single stage in the N level circuit could in principle implement an arbitrary boolean switching function by an appropriate arrangement of switches. This might require an exponential number of switches and so multiple stages appear to be useful. Also, although for simplicity we have assumed that the outputs of stage i are used directly only in stage i+1, they could equally be used in stages i+2, i+3, etc. We will not consider these additional complexities here because they are secondary to the fundamental issues.

LATCHES

What has been described so far will allow the construction of an N level reversible combinational logic circuit. To be useful, we must also specify a latch which can store the output of the circuit. This can be done by holding charge on a capacitor. The outputs of stage N are connected through switches to a latch stage, stage N+1.

The latch requires its own clock. Such a latch is shown in figure. Initially, the latch clock is 0 and allows charge to move freely into the capacitor. When the output from stage N is valid, the latch clock goes high (changes to logic 1) and prevents charge from escaping.

Clock N can now go low (to logic 0) which will make the output from stage N also go low, but the output of the latch will not be effected by changes in the output from stage N because it is cut off. Of course, having once "locked up" the outputs in the latches, we cannot release the information in the latches. If we should again set the clock latch low, allowing charge to flow off the capacitor, then we would be turning on a switch when there was a voltage across it. This is forbidden, and so the problem of removing information from a latch is non-trivial.

UNWRITING

Rather than simply "erasing" the information on the latch by letting the latch clock go low, we need to "unwrite" the information. Unwriting differs fundamentally from erasing. When we erase, we do not know the state of the latch, we simply turn on the switch holding the charge on the latch and let the charge flow away.

Whether the latch held a "0" or a "1" is immaterial. In the process, we must necessarily dissipate energy because erasing information is irreversible. By contrast, when we "unwrite" a latch we actually have an input signal that tells us the state of the latch. We can set the latch clock low and connect the input signal and the value stored on the latch.

Because (by definition) the input signal is in the same logical state as the latch, there is no voltage across the switch when we turn it on. We can now let the input signal go to 0 slowly, thus draining charge from the latch.We can indefinitely reduce the energy dissipated by this process by slowing the speed with which we drain charge from the latch.

Summary of Combinational Logic

In our earlier description, we left the source of valid input a mystery. The outputs of stage 0 held valid data, and how this happy state of affairs came about was unspecified. Now we can rectify this omission: stage 0 is a latch. This implies that clock 0 must be kept high to prevent the charges in stage 0 from leaking away.

Further, stage N+1 is a latch, and holds the output. We still leave the mechanism that initially put proper charge levels into the capacitors of stage 0 a mystery, but it is now clear that stage 0 will continue to hold a valid signal until clock 0 is lowered and the charge is allowed to escape.

The latches of stage 0 will also be called the input latches or simply the input, while the latches of stage N+1 will also be called the output latches or simply the output.

We can now specify our initial state as one in which appropriate charges are held on the input, clock 0 is high (to prevent charge from the input latches from leaking away), while clocks 1 through N are held low; as is clock signal N+1 (that is, there are no charges held on the output latches and the switch connecting the output latches to the final logic stage is turned on).

We raise clock signals 1 through N in turn, computing the values of each logic level in turn. Finally, we raise clock signal N+1, thus trapping the output signals in the output latches. We are then free to lower clock signal N, then N–1, then N–2, etc. until we have finally worked our way back to the first logic level. When the computation is done, all clock signals from 1 through N are held low, while clock signals 0 and N+1 remain high. Charge is trapped in stage 0 corresponding to the input, and charge is trapped in stage N+1 corresponding to the output. The outputs of stage 0 and N+1 are valid. All other outputs are 0.

ITERATION OF A REVERSIBLE LOGIC FUNCTION

At a higher level of abstraction, we can say that the outputs of stage N+1 are equal to some combinational function F applied to the outputs of stage 0, or output = F(input). Of course, we're now stuck. We cannot unwrite the input latches because we don't know what information is stored in them. We can solve this dilemma by requiring that F be reversible. That is, we demand that F^–1(F(input)) = input for all possible values of the input on stage 0. This allows us to compute the values stored in stage 0 from the values stored in stage N+1. We need merely compute F^–1(output).

That is, we add additional logic stages N+2, N+3..., N+M which compute the inverse of F. The output of stage N+M will then be the same as the input. Because we now have data values which are identical to the contents of stage 0, we can unwrite stage 0 in a fully reversible fashion.

That is, we can lower clock 0 and allow the charge on the capacitors in stage 0 to be connected with the output signals from stage N+M. Because they are identical the energy loss from this operation can be made arbitrarily small.

Then, clock signal N+M can be lowered, clock signal N+M–1 can be lowered, etc. until we finally lower clock signal N+2. Clock signal N+1 (the latch control clock signal for the output) is high (to hold the output steady in the output latches) and remains high.

To review the cycle of operations from the initial state: we start with a valid input in stage 0, with clock signal 0 held high (to hold the charge in the latches of stage 0) and with all other clock signals held low. From this intial state, we start the computation by raising clock signals 1 through N in turn, computing F(input). Clock signal N+1 is then raised to hold the output values in the output latches of stage N+1, and then clock signals N, N–1, N–2..., 1 are lowered in their turn to "uncompute" the calculation of F. We then compute F^-1(output) by raising clock signals N+2, N+3, N+4..., N+M. This computation produces exactly the values that are stored in stage 0, and so we can now unwrite the contents of stage 0 by lowering clock 0: this turns on the switches connecting the latches of stage 0 to the output of stage N+M.

We then uncompute the calculation of F^{-1}(output) by succesively lowering clock signals N+M, N+M–1, N+M–2..., and N+2. Finally, we again raise clock 0 to again isolate the input (which is now all 0). At the end of this process the outputs are stored in stage N+1 and the input is all 0. The entire process was carried out in a fully reversible manner, and so energy dissipation can in principle be made arbitrarily small.

The only thing we need finally do is exchange the data in the input and output latches (we assume that the input and output latches have the same number of bits). To do this, we first transfer the data in the output latches into the input latches and then "unwrite" the output latches. While the circuitry for this is not entirely trivial, it should be clear that the principles developed should allow us to do this. The sequence of steps involved is gone over in detail in the following several paragraphs.

To exchange the data between the input latches and the output latches, we require four additional sets of switches and four additional sets of clock lines. We illustrate the process by considering the exchange between a single input latch and a single output latch. Initially, EC_1 and EC_2 are set to 0 while EC_3 and EC_4 are set to 1. The switches S_3 and S_4 are turned off. S_1 is turned on, because the input latch holds a 0. S_2 could be either on or off, depending on the data on the output latch.

We first transfer data from the output latch to the input latch. EC_3 is lowered turning on S_3. Because the input latch holds 0 and EC_2 is set to 0, turning on S_3 is permitted. EC_2 is then raised to 1. If S_2 is on, charge is transferred to the input latch. If S_2 is off, charge is not transferred.

Thus, if the output latch holds a 1 the input latch will be set to 0. If the output latch holds a 0, the input latch will be set to 1. In the process of transferring the data from the output latch to the input latch, we have also inverted it. To correct this omission would require an inverter in the diagram of figure, or some modifications to F and F^{-1}. This would take an additional switch and an additional clock line, as well as additional explanation. The figure, however, would rapidly become cluttered so we have chosen to omit the inverter to clarify the exposition. This ommision has no impact on the basic concept.

Having set the input latch with data from the output latch, we now raise EC_3 and turn off S_3. Finally, we lower EC_2. This completes the transfer of data from the output latches to the input latches. We must now unwrite the output latches. Unwriting the output latches is similar to setting the input latches, but the exact sequence of operations is changed. Our first step is to raise EC_1. We then lower EC_4 turning on S_4. EC_1 is high, so if there is charge on the output latch then we have turned on a switch when there is no voltage across it, which is permitted.

If there is no charge on the output latch, then there will be charge on the input latch (because we transferred the complement of the output latch to the input latch in the earlier steps) and this charge will turn off S_1. Therefore, the output latch and EC_1 will not be connected. Again, we have not turned on a switch with a voltage across it.

At this point, if the output latch is charged it is connected to EC_1. By lowering EC_1 we can drain the charge from the output latch. We then raise

EC_4 turning off S_4. This isolates the output latch again and completes the unwriting of the output latch from the input latch.

The clocking sequence, in brief, is: lower EC_3, raise EC_2, raise EC_3, lower EC_2, raise EC_1, lower EC_4, lower EC_1, raise EC_4. This sequence of operations has exchanged the data on the input latch and the output latch, and did not require turning on a switch when there was a voltage across it. The careful reader might wonder about the charge that might be trapped on the segment between S_1 and S_4 or between S_2 and S_3. These segments are initially 0, and are restored to 0 upon completion of all operations.

During the actual transfer and unwriting steps, the charge on these segments is always known and does not result in turning on a switch with a voltage across it. An explicit capacitance attached to this segment would not alter the conclusions reached here for this specific design.

In other designs it might be useful to eliminate such segments. Whether or not such segments can be eliminated depends on the specific implementation. Elimination can often be done by appropriate physical design of the switches, effectively resulting in a single switch with two control inputs.

We have now completed the full cycle of operations required to compute F(input). We can go on to compute F(F(input)), F(F(F(input))), etc. We assume F is computationally useful (*e.g.*, it implements a single instruction or step of a reversible computation). As discussed earlier, reversible functions can be used to carry out arbitrary computations.

PROBLEMS WITH INACCURATE VOLTAGE

While the system described is reversible in principle, the energy dissipation in a practical implementation would depend upon the accuracy of the voltage. When a switch is turned on, the logical state of the input and output values is the same, but the physical state might differ.

A small difference in the voltages at the input and output of the switch would result in energy loss as the switch was turned on. Even with this restriction, it should be possible to achieve significant reductions in energy dissipation. The energy dissipated by dissipatively discharging a capacitive load is,

$$\text{Energy} = 1/2\ CV^2$$

where C is the capacitance and V is the voltage. In conventional (irreversible) logic systems, the voltage V would be the voltage of the power supply. In a reversible system of the type described here the voltage V would be the error in the voltage of the power supply, rather than the voltage of the power supply itself.

Thus, by using reversible instead of irreversible logic, we have effectively reduced the value of V. With a power supply accurate to 5 per cent, the energy losses from charging and discharging the load capacitances associated with the electrical connections would be decreased by a factor of 20^2 or 400. That is, energy losses would be a function of the error in the voltage of the power supply, rather than a direct function of the voltage of the power supply itself.

There does not appear to be any reason in principle why the error in the voltage of the power supply could not be made as small as desired. A more fundamental reason for long term optimism is the observation that electric charge is quantized. In future systems in which the number of electrons that represents a logic "0" or "1" is both finite and small, it should be possible to reduce the energy loss caused by slight inaccuracies in the voltage supply to 0.

It is already possible to control the passage of individual electrons in a circuit. Such highly precise control should eventually make a fully reversible electronic switch with extremely small energy losses feasible.

Problems with FETS: Voltage Degradation

Switching elements that do not degrade the voltage of the signal being switched are feasible in principle, as shown by the electrostatic relay illustrated in figure. CMOS transmission gates can also be used as switches without concern about voltage degradation. Transmission gates require that both the control signal and its inverse be provided. A simple method of satisfying this requirement would be to use two-rail logic. To use nMOS (or pMOS) by itself, we must overcome certain shortcomings.

In particular an nMOS switch, when used in pass logic, will cause voltage degradation. As a consequence, deterioration of the voltage through multiple logic levels would impose a severe limit on the capabilities of the system.

Boosting voltage in a fully reversible fashion is, however, straightforwards in principle. The voltage on a capacitor will increase if the capacitance decreases. Variable capacitors are well known based both on physical motion of the capacitor plates, or on changes in charge distribution secondary to a voltage change. Varactors in which the capacitance C changes by a factor of 10 or more in response to a voltage change of little more than a volt are feasible. The approach outlined here is abstractly similar to the proposal by Joynson et. al.

Boosting voltage is illustrated in figure. Before starting, the switch is turned on and current can flow onto the varactor. The high-voltage clock then opens the switch, trapping the charge on the varactor. Finally, the capacitance of the varactor is reduced. Because the voltage on a capacitor is $V=Q/C$ (where V is voltage, Q is charge and C is capacitance) the voltage across the varactor will increase as its capacitance decreases.

Because the voltage on the high-voltage clock is large, there is more than sufficient control voltage to insure that the switch remains turned off even though the voltage across the switch has been increased. The low-voltage input, boosted to a medium-voltage output, can then be used to switch a low-voltage signal, which can again be used as the input to a reversible voltage booster. To reverse this cycle of operations, the capacitance of the varactor is returned to its original value.

This will also return the voltage on the output to its original value, and the high-voltage switch can be turned on with minimal energy loss. This energy loss will be a function of the square of the error in the voltage regulation. If the number of electrons that charge the varactor (and any stray output capacitance) is reduced to a small integer number, it should eventually be possible to eliminate losses due to small inaccuracies in voltage regulation. Seitz et. al. commented that "...the isolation transistor can turn on while there is voltage across it, and accordingly, it dissipates power in charging or discharging the bootstrap node. The goal of exporting all of the dynamic power is elusive." Translating from their language to the language used here, they sometimes allowed the high-voltage clock to turn on the high-voltage switch when a voltage existed across the switch, thus dissipating energy.

They did not consider the computation of F ^ -1 and "unwriting" operations, and so were sometimes obliged to engage in the dissipative operation of discharging the gate of a FET to ground. In essence, they were able to eliminate unnecessarily dissipative operations for combinational logic, but not for sequential logic. As shown here, appropriate design of a reversible sequential circuit eliminates the unnecessarily dissipative steps.

PROBLEMS WITH FETS

Leakage Current

Another problem with FET's is leakage current. Even a FET in which the gate voltage is sufficient to insure that the FET is completely cut off will still pass a small current. If our objective is to achieve energy dissipations below kT, any significant leakage current is intolerable. Leakage current occurs because thermally generated electron-hole pairs are created in the semiconductor material of the FET.

Thermal agitation will occasionally cause an electron in the valence band (*e.g.*, an electron which normally would not be able to move because it was fixed in place as part of a covalent bond) to leap into the conduction band. An electron in the conduction band is free to move throughout the crystal and so to conduct electric current.

The energy gap between the valence band and the conduction band will govern the concentration of electrons that are available to carry leakage current. In silicon, the bandgap is 1.12 electron volts.

At 300 kelvins, this will produce 1.45×10^{10} electrons per cubic centimeter of silicon. This "intrinsic carrier concentration" will result in dissipative energy losses. Three obvious methods are available for reducing the leakage current:

- *Reduce the physical size of the device*: An intrinsic carrier concentration of 1.45×10^{10} electrons per cubic centimeter can also be expressed as 1.45×10^{-11} electrons per cubic nanometer. If the active region of the device was smaller than a few millions of cubic nanometers, then the probability of finding an undesired charge carrier in that region would be relatively small.

- *Reduce the temperature*: *The intrinsic carrier concentration is*:

$$ni = km \exp(-Eg/2kt)$$

where ni is the intrinsic carrier concentration, km is a constant specific to the material, and Eg is the bandgap energy. In the case of silicon, km is about 3.1×10^{20}. At a temperature of 77 Kelvins, ni is 7.4×10^{-17} electrons per cubic centimeter. CCD's with dark currents of a few electrons per pixel per hour have been fabricated.

- Select a material with a larger bandgap. Diamond has a bandgap of 5.47 electron volts, with an intrinsic carrier concentration at room temperature in the range of 10^{-26} electrons per cubic centimeter. Boron provides a good shallow dopant, and p-type diamond has been made. Several shallow dopants for n-type material are plausible. Diamond MOSFETs with boron doping have been fabricated. An implementation could use purely p-type materials, if that proved useful.

  A variety of wide band-gap materials exist allowing a wide range of tradeoffs between energy dissipation and other parameters. It should be feasible to produce acceptable leakage currents using a variety of different materials in a variety of different geometries under a broad range of operating temperatures. The use of small FET's using a wide band-gap material is particularly attractive for room temperature operation. In such a structure the probability of finding even one valence electron in the conduction band is astronomically small.

Thermally generated leakage current is also of concern in many types of varactors. A reverse- biased P-N junction illustrates the problem: electron-hole pairs generated thermally in the charge depletion region are swept out of the region, thus conducting current. Leakage current in such varactors can in principle be made quite small for the same reasons that apply to leakage current in a FET.

Reversible CCD-Based Logic

An alternative approach, which eliminates the need for high accuracy voltage supplies, is to eliminate the need to turn on the switch when there are charge carriers present. This might be viewed as only turning the switch on and off when the voltage across the switch is 0 volts, but this is not a fully adequate description. Even a 0 volt signal is subject to fluctuation and noise. If, however, we try to put a positive voltage onto the switch (*e.g.*, the inputs and outputs of the switch both have a positive voltage), and if the charge carriers in the switch are electrons, then it is possible to remove all the charge carriers from the channel region.

Any charge carriers will be actively swept out of the channel region by the positive voltages on the input and output leads. Once all the charge carriers have been removed from the channel region, then no more current will flow. Precise regulation of the positive voltage on the input and output of the switch is not required, for both need merely be "sufficiently positive" to insure removal of all charge carriers from the channel region. Once the charge carriers have been removed then turning the switch on and off can be done in a fully reversible manner. Small errors in the supply voltage can no longer lead to dissipative current flows when the switch is turned on, because no current can flow.

For purposes of this discussion we will assume our switch is a FET with electrons as the charge carriers and that the electrons are removed from the FET by a positive voltage on the input and output leads. (We could equally well have assumed the charge carriers are holes and that a negative voltage is placed on the input and output leads: for purposes of explanation the choice is arbitrary).

As discussed earlier, it is essential that the device be designed so that the concentration of intrinsic charge carriers is sufficiently low that energy losses from this source can be neglected.

Further, the density of device defects must be low enough that defect sites do not appreciably alter device performance.

We will simply assume that the material is free of defects. Reduction of device size coupled with decreases in defect rates should eventually allow construction of devices with a sufficiently low defect rate that energy dissipation will not be significantly degraded.

In this approach, 1's and 0's are represented by a packet of electrons or the absence of a packet of electrons, respectively. The electrons are stored in potential wells which we will call "buckets." Each packet of electrons is treated as a unit: we never split or merge packets, nor do we create or destroy packets (except when the system is initialized, which would ideally occur only once). While all packets have approximately the same number of electrons, it is not critical that each packet have exactly the right number of electrons. Whether a packet has 99, 100, or 101 electrons won't matter.

The use of charge packets is most familiar in the context of CCDs (Charge Coupled Devices), or alternatively in the context of BBD's, or Bucket Brigade Devices. In a CCD, charge packets are transferred from potential well to potential well serially along a row of devices.

The amount of charge in each packet is a measure of some analog quantity. Such devices are primarily used as memory elements or shift registers, not for computation. Proposals to use the same basic concepts for logic operations have been made since the invention of the CCD.

Sometimes called "DCCL," or Digital Charge Coupled Logic, these proposals have explored the advantages of using charge packets as the basic logic element. The primitive logic elements in prior proposals, however, were highly dissipative. For example, Zimmerman et. al. proposed logic operations whose computation required merging packets, discarding packets by draining the charge to ground, etc.

The goal of thermodynamically reversible logic was not an objective. Tompsett proposed circuits in which the potential generated by a charge packet in a CCD potential well was coupled to the gate of a FET. This FET would then be turned on or off depending on whether a charge packet was present or absent in the potential well. This was then used to create another charge packet (rather than to control the movement of an existing charge packet). The creation of a new charge packet would compensate for imperfect charge transfer.

NAND and NOR operations could be performed by applying the potential from two potential wells to two gates of two FETs, which could then be connected in serial or parallel. Again, the objective of thermodynamic reversibility was not considered. The method of sensing charge in the potential well was dissipative and the sensed packets, no longer needed, were to be dissipatively discarded. For simplicity, we will assume that packets never gain or lose electrons (due to defects in the material or to thermal noise).

In fact, present day devices can gain or lose charge at the rate of a few parts per million. As discussed earlier, such gains or losses should eventually be reduced to insignificant levels. Even if we allow packets to gain or lose electrons at some low rate, we could periodically "refresh" the packets, and so prevent malfunction. Tompsett first proposed methods of refreshing packets (though he did not consider reversible methods). To the extent that the packet has gained or lost an unpredictable amount of charge, "refreshing" the packet must fundamentally dissipate energy. Other than this fundamental loss, however, we wish to keep energy dissipation to a minimum.

A reversible method of refreshing a packet would be to use the old packet to charge a bucket in a reversible fashion (as discussed in the first method), and then to "unwrite" or discharge the bucket holding the old packet by using the new packet as the source of information. This method of refreshing a packet would, of course, dissipate energy as a function of the errors in the voltages involved (as discussed earlier). If the refresh operations were done infrequently (*e.g.*, a packet would be refreshed only after a large number of logic operations) then the energy dissipation per logic operation from this source would be quite small.

While CCD and DCCL circuits are today implemented on a semiconductor surface using planar technology, for purposes of this discussion we will initially ignore this geometrical constraint and instead focus on more fundamental issues. A planar version of the ideas described here is given in the section on "A Planar Version of RCT Logic."

Illustrates the "hydraulic" model of charge transfer in a CCD. Charge is transferred from one potential well to the next much like water would flow to a gradually sinking region in a pond. This process can be thermodynamically reversible when appropriately implemented. One aspect of figure might be slightly misleading. While water has a sharply defined surface, the energy of an electron at some non-zero temperature T is statistical. While figure might lead one to conclude that extremely small perturbations in the potential might cause significant changes in the pattern of charge flow, in fact thermal noise will allow electrons to surmount small barriers *e.g.*, where the barriers have a potential similar in magnitude to kT/e or.026 volts at room temperature (where e is the charge on an electron). Small perturbations and errors in the potential do not fundamentally create dissipative problems.

TRANFERRING CHARGE WITH LITTLE DISSIPATION

Before discussing the more complex operations needed for computation, we first consider a simple sequence that transfers charge from a source bucket to a destination bucket with as little energy loss as desired. This is illustrated in figure. (Note that the illustration shows that the two buckets have semiconductor plates joined by a semiconductor path. The use of metallic plates or a metallic connection between the two plates would cause unwanted energy dissipation). At first, both the source clock and destination clock are + (positive), and the charge packet is entirely in the source bucket.

The clocking sequence will then shift the charge from the source bucket to the destination bucket in a series of steps, where each step can dissipate an arbitrarily small amount of energy. As we move from state 1 to state 2, the destination clock is changed from + to -. This change results in negligible energy dissipation, because no charge can move at this time. As we move from state 2 to state 3, the source clock is changed from + to -. The charge carriers in the charge packet (electrons in our example) now move out of the source bucket. Because the destination clock is negative, the electrons will migrate smoothly out of the source bucket. (If the destination clock were + at this point instead of -, the electrons would move out of the source bucket and then "fall down hill" into the destination bucket, dissipating energy). As we move from state 3 to state 4, the destination clock changes from-to +, thus slowly attracting the electrons and gathering them into the destination bucket. Finally, as we move from state 4 to state 5, the source clock changes from-to + restoring the system to its original condition, but with the charge packet in the destination bucket instead of the source bucket.

The primary source of energy dissipation in this sequence is from state 2 to state 3, and from state 3 to state 4. These transitions involve charge migration, and will therefore result in dissipative losses. As we slow the clock frequency the clock voltages change more slowly, the charges move more slowly, and hence dissipate less energy. Because energy dissipation is a function of the square of the current, reducing the frequency of operation by a factor of two will also reduce the current by a factor of two but will reduce the energy dissipation by a factor of 4. This will reduce the energy dissipated per logic operation by a factor of 2. Thus, by sufficiently slowing the clock frequency we can dissipate as little energy per operation as might be desired. It is important to notice that these operations do not erase information. We start with a charge packet in a single potential well on the left, and we end with a charge packet in a single potential well on the right.

During the course of moving the charge packet we slowly change the shape and position of the potential well, but at no point do we merge potential wells. This transition eliminates the right hand potential well without merging it with the well on the left. An attempt to merge the two potential wells would be dissipative. The transition from step 2 to step 3 slowly changes the shape of the potential well, but the charge packet is always contained in a single potential well and is always at or near thermodynamic equilibrium within that well.

THE PRIMITIVE OPERATION

We now consider the somewhat more complex operation that is required to support computation in a reversible fashion.

The single primitive operation is: Transfer charge out of a source bucket and into an empty destination bucket if a "condition" bucket holds a "0", but do not transfer the charge if the condition bucket holds a "1". If the transfer does not take place (the condition bucket is "1"), the destination bucket need not be empty. "Empty" in this context means "has no charge carriers present." That is, an "empty" bucket is in logic state "0." We will call this device a 3-bucket Reversible Charge Transfer device, or (where context makes it clear which type of RCT we are referring to) simply an RCT.

The RCT is a reversible DCCL device. At a logical level the RCT is related to (but somewhat different from) a switch gate implemented using charge packets instead of colliding billiard balls. The particular logic function described here is but one example of the range of logic functions that involve the reversible transfer of charge from one bucket to another.

During normal operation, RCT devices:

- Never merge or split charge packets
- Never discharge charge packets to ground, and
- Move charge packets in a thermodynamically reversible fashion from a set of source buckets to a set of destination buckets, conditional on the presence or absence of charge in a set of condition buckets.

In this context thermodynamic reversibility means that energy dissipation per device operation, in the absence of manufacturing defects, could be made much smaller than kT simply by slowing the speed of operation.

The number of charge packets is conserved and therefore RCT devices are conservative logic devices (in conservative logic devices, the number of 1's is conserved during the computation, so the number of 1's at the input is the same as the number of 1's at the output).

We have described the 3-bucket RCT, which has one condition bucket, one source bucket, and one destination bucket. More complex RCT's have more buckets and the pattern of charge transfer is more complex. Some of these RCT's will be considered later.

The simple 3-bucket RCT operation might be written in C as:

- If (Condition==0) {Destination = Source; Source = 0}; while a somewhat briefer notation would be:
- If Condition is 0 then Destination = Source. (Note that in this briefer notation, the "=" sign is used to indicate both the assignment of the Source value to the Destination, and setting the Source to 0 after the assignment).

The precondition that must be true prior to execution of the 3-bucket RCT is:

(Destination==0) OR (Condition==1)

Violation of this precondition would result in an unwanted dissipative step, which is banned in RCT devices. While it might at first seem that satisyfing this precondition would make the 3- bucket RCT data dependent, it is in fact possible to design a circuit in which we know that the precondition is satisfied without knowing what values are presented as inputs to the circuit. A logic diagram showing the possible initial and final states of a 3-bucket RCT is shown in Figure.

On the left, the legal initial states of the RCT are shown, while on the right the final states (after the sequence of clock signals that cause the device to step through a single cycle of operations) are shown.

It is interesting to note that there is only a single legal, non-trivial change of state: when the input is 1, the condition is 0, and the output is 0. In all other cases either the input and output states are the same or the input state is illegal. The illegal states would cause undesired energy dissipation.

FREDKIN GATE WITH 3-BUCKET RCT

The 3-bucket RCT is logically complete, for we can implement a Fredkin gate with it. The construction used here is related to (though diffrent from) that used by Fredkin and Toffoli for constructing a billiard ball Fredkin gate from a billiard ball switch gate.

A Fredkin gate has three inputs and three outputs. The inputs are the control input A, and two signal inputs B and C. The outputs are A', B', and C'. If the control input A is 0, then A' = A, B' = B, and C' = C. If the control input A is 1, then A' = A, B' = C, and C' = B. That is, the output of the Fredkin gate is identical to the input if A = 0, but B' and C' are exchanged if A = 1.

The following sequence of simple RCT operations will implement a Fredkin gate: A, B, and C hold arbitrary input values. InitiallyOne is set to a logic "1" (the presence of an electron packet). All other variables are initialized to logical "0" (the absence of an electron packet).

- If A is 0 then C' = C.
- If A is 0 then B' = B.
- If A is 0 then NotA = InitiallyOne.
- If NotA is 0 then C' = B.
- If NotA is 0 then B' = C.
- If AlwaysZero is 0 then A' = A.
- If A' is 0 then InitiallyOne = NotA.

Steps 1 and 2 simply transfer charge from B and C into B' and C' if A is 0. B' and C' are initially empty, so this transfer satisfies the RCT preconditions.

Step 3 computes the logical negation of A and leaves that logical negation in "NotA."

Steps 4 and 5 transfer charge from B and C into C' and B', exchanging the outputs. The precondition is satisfied, for if A is 0, then NotA will be 1 and the transfer will not take place. If the transfer does not take place, the prior contents of B' and C' are irrelevant.

If A is 1, then NotA will be 0 and the transfer will take place. In this case, B' and C' will be empty (have logic values 0) and so the transfer will take place correctly.

Step 6 copies A into A' unconditionally.

Step 7 restores the value of InitiallyOne if it was altered during the computation.

This implementation of a Fredkin gate will take three arbitrary input values and, after a sequence of RCT operations, produce three output values. We could clearly apply N Fredkin gates to 3N input values and produce 3N output values at the same time. We require that all output values be initially 0. After finishing the sequence of clocking operations, the inputs are all 0 and the outputs hold arbitrary values. At this point, the "input" and "output" can be logically exchanged (*e.g.*, no physical operation takes place, but we relabel the "input" variables as "output," and the "output" variables as "input"). Following this logical exchange, we have reestablished the preconditions for the next sequence of operations: the (just relabeled) inputs again hold arbitrary logic values, while the (just relabeled) outputs hold logic 0's.

We now drive a (different) set of clock lines to apply the (different) Fredkin gates to generate the next "state" of the computation. Viewed globally, this is a "ping-pong" or "double buffer" scheme for computing successive values of the global state of the system. If we call the two sets of variables A and B, and we call the two logic functions for the "next state" of the system F and G, then we first compute B=F(A). B is initially all zeroes, while A holds arbitrary logic values.

On completion of the computation of F, A is all zeroes while B holds arbitrary logic values. We then compute A=G(B). Because G uses B as its input and A as its output, the condition that the outputs be all zeroes is met during the compuation of G. This returns us to the state where A holds arbitrary logic values and B holds all zeroes. We can repeat this cycle indefinitely.

IMPLEMENTATION OF A 3-BUCKET RCT

A planar layout of a 3-bucket RCT is shown in figure. Many kinds of implementations would be feasible, for example Silicon-On-Insulator technology (the FET used for the transfer switch would be "tilted on its side" in this approach). Note that many of the conductive paths are really semiconductors. Each bucket also has one semiconductor plate. The gate is also a semiconductor. Use of metallic wires for these components would compromise device function. A planar RCT is described in a later section.

This RCT can be viewed as a combination of three CCD potential wells and a FET. A charge packet is first transferred from the "condition" potential well to the gate of the FET. The charge packet in the "source" potential well is then moved to the source of the FET and (the charge on the gate of the FET permitting) through the channel and out the drain of the FET into the "destination" potential well.

Tompsett proposed sensing the presence or absence of a charge in a CCD potential well by connecting the potential well via a doped region connected ohmically to a metalization that became the gate electrode of a FET. Charge injection from the gate into the doped region, while small if the capacitance of the gate is small, still introduces an unwanted source of energy dissipation. Other aspects of Tompsett's proposal were also irreversible or highly dissipative. In the present proposal, we require that both the gate of the FET and the connecting path from the potential well to the gate be a semiconductor. The only charge carriers in this semiconductor path would be those deliberately introduced as a charge packet.

An alternative approach which might be more convenient in a simple planar implementation would be to introduce a capacitive coupling into the path from the potential well to the gate. The gate could then be metallic, one plate of the capacitor (the plate connected to the gate) could also be metallic, but the other plate and the connection to the potential well could be a semiconductor.

Potential could be coupled through the capacitor, but charge carriers would be prevented from moving into or out of the semiconductive region and hence charge injection from the gate electrode would not be a problem. Initially, a charge of unknown value is present in the source bucket. The destination bucket is either empty (if the condition bucket holds a 0) or has an unknown charge (if the condition bucket holds a 1). Clocks 1, 2, and 3 are all positive. All charge carriers (electrons) have been drawn onto the semiconductor plates of the buckets.

The conventional plates are connected to the clocks. There is a net positive charge on each bucket creating a potential well which holds the electrons in the bucket and prevents them from migrating away. The channel regions of all switches have no charge carriers present. Clock 4 is driven negative with a large voltage.

This large voltage cuts off and isolates the three buckets involved from any other buckets to which they might normally be connected. Additional switches controlled by clock 4 might be required to further isolate the three buckets in a larger circuit, but are not shown in this small example.

The two gray channels do not actively participate in the process of reversible charge transfer, they are present merely to show that connections to other circuit components will be present in a real system.

Clock 3 is driven negative, forcing charge from the condition bucket onto the gate of the transfer switch. If no charge is present on the condition bucket, no charge will be put onto the transfer gate, while if charge is present then it will be transferred to the transfer gate, cutting off the transfer switch.

Note that the circuit is so arranged that only a medium voltage will be developed on the transfer gate, and this medium voltage can be contained by the high voltage of the cut off clock, clock 4.

Clock 2 is driven negative. If there is charge on the destination bucket, then the transfer switch is turned off and the charge on the destination bucket goes nowhere. If there is no charge on the destination bucket, then whether the transfer switch is turned off or on is immaterial, for no charge will be transferred.

Clock 1 is driven negative. If the transfer switch is turned off, no charge will be transferred. If the transfer switch is turned on, charge will move gradually through the transfer switch into the destination bucket. Because clock 2 is negative, charge will not "fall down hill" in a dissipative fashion when moving into the destination bucket.

Clock 1 has a low voltage so a low voltage is generated during the transfer operation. The medium voltage on the transfer gate is sufficient to block charge transfer if charge transfer is not supposed to take place.

Clock 2 is driven positive, allowing charge to flow smoothly into the destination bucket. When clock 2 reaches its full positive value, all the charge is held on the destination bucket and no charge carriers are present in either the source bucket or the transfer switch channel region (if the transfer switch was turned on).

Clock 1 is driven positive, which has no effect if the transfer took place and simply allows the charge to return to its original state if the transfer did not take place.

Clock 3 is driven positive, allowing the charge on the transfer gate to return to the condition bucket.

Clock 4 is driven positive, opening the cut off switches that isolated the three buckets from outside influence during the reversible transfer operation.

This cycle of operations reversibly transfers charge from the source bucket to the destination bucket if the condition bucket holds a logic "0" (*e.g.*, no charge).

Energy Loss

There are losses during the RCT cycle of operations caused by current flowing through a resistive medium. The power lost will be, Power Loss = I^2 R where I is the current and R the resistance. Current is proportional to the frequency of operation of the circuit, so if the frequency is low the current will also be low. Therefore, power losses from resistive losses will fall off with the square of the frequency: Power Loss is proportional to f^2. Because the number of operations per second is also lower at a lower frequency, the actual energy loss per device operation is proportional to the frequency of operation: Energy Loss is proportional to f. By reducing the frequency of operation, the energy loss per operation can be reduced to whatever extent is desired. It should be remembered that "low frequency" operation is relative: CCD's with 0.9997 charge transfer efficiency operating at 1 gigahertz have been demonstrated.

More specifically, the electron drift velocity vd equals the electric field E times the mobility m:

$$vd = Em$$

If we let d be the distance traveled by an electron as it moves from one bucket to the next, and t be the time of a single logic operation, then it is approximately the case that:

$$d = vd\ t$$

The total energy dissipated by the movement of n electrons as they move from the source to the destination is simply the force times the distance times the number of electrons, or:

$$Edissipated\text{-}neEd$$

where Edissipated is the energy dissipated, n is the number of electrons in a packet, and e is the charge of an electron.

Putting these together yields:

$$Edissipated = ned^2/mt$$

If we assume that the distance traveled by the charge packet as it moves from one bucket to the next is.1 microns (10^{-5} centimeters), that one packet has 100 electrons, that the mobility is 1800 cm^2/Vs (the approximate mobility of diamond or silicon near room temperature), that the time allotted for one operation is 1 nanosecond, and the charge e on an electron is 1.6×10^{-19} coulombs, then the energy dissipated is approximately 10^{-21} joules.

This estimate is based on a classical model and involves some significant simplifications. Most significantly, the mean free path will of necessity be less than.1 micron in a device whose maximum dimension is.1 micron, and so the meaning of the electron mobility m is somewhat obscured.

A more detailed quantum analysis is essential as the size becomes smaller. A rough approximation to such an analysis can be based on the observation that the resistance of the channel of a sufficiently small FET (*e.g.*, one in which the channel width is perhaps a few nanometers and only a single transmission mode exists) is roughly half of h/e^2, where h is Planck's constant.

This is a fundamental unit of resistance of about 26K ohms. A packet of n electrons flowing in time t produces a current of ne/t. The power produced by such a current flow is $I^2 \times R$, while the energy dissipation per logic operation is $I^2 \times R \times t$, or $(ne/t)2 \times h/(2e^2) \times t$ which simplifies to:

$$\text{Edissipated} = hn^2/2t$$

For a switching time t of 1 nanosecond and a packet size n of 100 electrons, this is 3×10^{-21} joules for each packet that is passed through the channel of the FET. Under the specific conditions considered here this second approximation agrees at least in order of magnitude with the estimate based on a classical model. It is interesting to note that the two equations have significant differences. The resistance of the channel in the classical case varies with the length of the channel. The length of the channel is less significant in the non-classical case. The size of the charge packet, n, also influences energy dissipation in different ways in the two cases. In the classical case, the energy dissipation is a function of n rather than n^2.

In the classical case, two electrons drifting under the influence of the same electric field in separate regions of the channel will dissipate twice as much energy as a single electron. In effect, the resistance has been halved because the number of charge carriers has been doubled.

A similar effect would occur in a doped semiconductor if the doping density were doubled. If we assume a narrow channel that can only support a single transmission mode, however, then doubling the size of the charge packet will cause more electrons to move through the same channel in the same time period, with a resulting increase in interactions among the electrons. The reader should note that no claim of universality or high accuracy is implied for either of the above equations. They are rough approximations that apply only under a limited range of conditions.

These conditions need not, in general, be true. In particular, it would seem attractive to design devices in which single electrons are always close to their ground state both during charge transport and during switching operations.

This would occur if the electron were always confined in a time varying potential well of relatively small dimensions. Devices with this property should be feasible when we are able to manufacture devices of sufficiently small size. This should have a favourable influence on energy dissipation. The major factor ignored in this analysis is the energy dissipated by the clock signals. Many methods of providing periodic clocking pulses are possible. Three methods are discussed in the next section.

The mobility can be substantially increased by reducing the temperature. Mobilities greater than 1,000,000 have been demonstrated in GaAs-AlGaAs HFET's at around 10 K. While the impact of this is unclear when the device size is much smaller than the mean free path (which occurs with such high mobilities), it does mean that larger devices could be made which would have very low energy dissipation. The energy dissipation for this type of reversible electronic logic (as estimated here) is greater than the dissipation of Drexler's rod logic operated at a cycle time of 100 picoseconds (some 10 times faster than the 1 nanosecond considered here for the RCT). When RCT device operation using single electrons becomes feasible, then energy dissipation using RCT's would presumably be reduced.

Likharev's parametric quantron has an energy dissipation which (when adjusted for the temperature difference and the differing speeds of operation) is not as good as rod logic. The parametric quantron also requires low temperature.

Rod logic should work at room temperature. It is possible that mechanical molecular logic might have very favourable energy dissipation properties at low temperature. While it is premature to draw conclusions it is not obvious that electronic devices must necessarily prove better than molecular mechanical devices in terms of energy dissipation at a given speed and temperature of operation.

CLOCKING

To insure that overall energy dissipation is reduced we need a method of providing suitable clocks that does not itself dissipate too much energy. We first consider a conceptually simple method of providing arbitrarily complex clock signals which can be made to dissipate as little energy as desired.

The loads presented to the clock in the proposals discussed here are purely capacitive. If we assume, for the moment, that the energy dissipated when a clock signal charges and discharges such purely capacitive loads is $1/2\ CV^2$, then the only method of reducing the energy dissipation is to reduce the voltage. Assume, for the moment, that we want a clock signal to increase from 0 volts to 5 volts. If we simply connect the clock line to a 5 volt source, then the energy dissipated will be $1/2\ CV^2 = 12.5\ C$ joules.

If we have 10 voltage supplies, with voltages of 0.5, 1.0, 1.5, 2.0, 2.5, 3.0, 3.5, 4.0, 4.5, and 5.0 volts, then we can successively connect our clock line to each voltage supply in turn. The voltage to the clock line would then be stepped up in 0.5 volt increments, and so the energy dissipated by a single step would be $1/2\ C(V/10)2 = 0.125\ C$ (*e.g.*, the energy dissipation per step is decreased by a factor of 100 because the size of the voltage step is decreased by a factor of 10).

To increase the voltage from 0 to 5 volts involves 10 steps, and so we will dissipate 10 × (0.125 C) or 1.25 C joules. This is 10 times smaller than the 12.5 C joules we would have dissipated using a single 5 volt supply. Thus, be using 10 DC power supplies and switching the clock line successively from one power supply to the next, we have reduced energy dissipation by a factor of 10. Clearly, we can reduce energy dissipation by a factor of N if we use N DC power supplies.

Notice that the larger we make N the more time we must take to increase the voltage from 0 to 5 volts. Each time we take a step in voltage we must wait somewhat longer than one RC time constant for the voltage to settle at the new value. This settling time will remain about the same for each step, even as we decrease the size of the voltage step by increasing N. As N increases and the number of steps increases, the total time taken for N steps will also increase.

This approach is very flexible because the clock waveform can rise and fall at arbitrary times. Complex clock signals can be easily generated. There are many alternative clocking schemes. While the method just described is asymptotically non-dissipative if a sufficiently large number of DC power supplies are used, it still involves moving electrons through wires to carry the clock signal throughout the circuit.

This does not appear to be the major limiting factor in energy dissipation when using today's planar technology. However, if in the future we use single electron logic devices, then moving large numbers of electrons over long distances to provide the clock signals would seem inappropriate.

There are approaches to clock distribution which avoid the need to move charge through wires. A conceptually simple approach is to provide clock distribution by the use of charges that are fixed to a rotating tube or disk. A facing disk or a smaller concentric drum would carry the logic circuitry.

As the disk spun the charges fixed to its surface would move, but this motion would not involve electrons flowing within a conductor.

The electric charges on the surface of the disk would then move past the logic elements on the facing surface of the opposing disk, providing the repetitive voltage changes required to clock the logic circuitry. In principle this method of moving charges would result in no conventional resistive losses at all because the charge would not be moving through a conductor.

There would still be radiative losses caused by the acceleration of the charged particles, and electrostatic attraction and repulsion would produce forces between the charges on the two disks which would lead to losses simply by alternately compressing and decompressing the material of the disk. While this approach might not be attractive with current technology, it might prove attractive at some point in the future if strong materials could be rotated at high speeds, thus providing a high interfacial velocity between the two opposing surfaces; and the pattern of charges on the surface could be both precisely controlled and have high resolution.

This should eventually be feasible, though whether or not it will prove competitive with other approaches is uncertain. Another method of providing a time-varying electric field would be to use a simple rotating electric field. Such a rotating electric field could then be used to clock the circuit. While the methods described here do not obviously lend themselves to such an approach, helical logic uses such rotating electric fields to drive thermodynamically reversible logic operations.

Charge is transported along a helix much as water is transported along an Archimedes screw.

Single electron versions of helical logic should eventually be feasible. Rotating electric fields can be provided by several methods, including cavity resonators or multiple LC oscillators connected to plates arranged in an appropriate pattern in space.

OTHER RCT'S

The 3-bucket RCT is not the only type of RCT that is useful. For example, if we use two condition buckets C_1 and C_2, and require that the charge packets in C_1 and C_2 be complementary (*e.g.*, C_1 is the negation of C_2); and if we use a single source bucket S and two destination buckets D_1 and D_2; then we can implement the primitive "Transfer charge from S to D_1 if C_1 is false, but from S to D_2 if C_1 is true." We will again need a pre-condition that effectively prevents the merging of two packets; we omit the precise statement of this precondition.

This 5-bucket primitive is logically complete (we leave this as an exercise for the reader) and has certain implementation advantages.

In particular, because both C_1 and its complement C_2 are available, we need only "steer" the charge in the source bucket to either of the destination buckets. We need not "bottle up" the charge on the source bucket in the event the transfer is not to take place (as occurs with the 3-bucket RCT). Because the charge packet in bucket S is guaranteed to move to some destination bucket (either D_1 or D_2), the actual voltage that must be generated to prevent the charge from moving in the wrong direction is smaller than the voltage required in the simple RCT to prevent the charge from transferring to its single destination bucket.

The same logic function could also be implemented in a somewhat different fashion by using a 4- bucket RCT with a single source bucket S, a single condition bucket C and two destination buckets D_1 and D_2. This is illustrated in Figure. In this alternative implementation, the charge in C would control a single FET between S and D_1, while S and D_2 would always be connected. The key to the conditional transfer is the timing of the clock lines for the two destinations D_1 and D_2.

In particular, we would transfer charge to D_1 through the FET first. During this transfer, D_2 would be maintained in an unreceptive state (the clock for D_2 would be negative). After transfer of charge to D_1 through the FET, the clock for D_2 would be made positive.

This shift would cause charge to transfer from S to D_2 only if the charge had not previously been transferred to D_1 through the FET. Thus, charge would be moved conditionally to either D_1 or D_2 depending on the condition C. In some sense, this is logically equivalent to the simple 3-bucket RCT operation followed by the transfer of the charge in S to the second destination bucket D_2.

A 4-bucket RCT with a single source bucket S, a single destination bucket D and two condition buckets C_1 and C_2 could be used to transfer charge only if both C_1 and C_2 were at logic 0. Arbitrary switching functions could be imposed between the source and destination buckets, and the number of condition buckets could be increased arbitrarily.

A Planar Version of RCT Logic

It would be advantageous to have a planar layout for an RCT that used only sine waves for clocking. This would be easy to implement in current planar technology, and the clock signals could be generated by LC oscillators. These constraints can be met. In particular, we start by considering a 3-phase CCD as illustrated in figure. The three clocks, F0, F120 and F240, represent three sine waves offset by 0, 120, and 240 degrees respectively. Charge packets will move from left to right as the three clock signals vary sinusoidally over time. The illustration shows a top view, the black squares being the metalization regions.

The metalization typically would be on top of SiO2, with the bulk Si beneath that. A profile view of charge transfer is given in figure. To provide a switching operation, we must provide some sort of choice for the charge packets moving along the CCD. This is illustrated in figure, where the source CCD shift register divides into two offspring.

To control the direction in which the charge packets are moved, we assume that two "condition" CCD shift registers provide the needed condition packets that will cause charge to select one offspring shift register over the other. Note that this device is just a 5-bucket RCT with a somewhat different geometry and clocking sequence.

It uses 3-phase sine wave clocks to move charge through the device. An additional fourth phase (F180) is required to control the conditional charge transfer from the single large source well (the large rectangle in the figure) to the two destination wells directly to its right. In this particular planar RCT, Condition1 and Condition2 are two CCD shift registers which hold complementary charges. If a potential well in the Condition1 shift register contains a charge packet, then the corresponding potential well in the Condition2 shift register does not (and vice versa).

The condition charge packet moves from left to right. When it reaches the well clocked by F180, the presence or absence of the charge packet changes the voltage on the corresponding gate electrode. There are two gate electrodes controlled by the two complementary charge packets. The gate electrodes are between the source potential well (the large rectangular metalization) and the two destination potential wells to its right. The gate electrodes determine the destination well into which the charge packet in the source potential well moves.

The voltage of the fourth phase, F180, is chosen so that the potential on the gate electrode is somewhat lower than the potential in the two adjacent potential wells when those two potentials are equal (*e.g.*, when F120 = F240)and when the condition charge packet is absent. Thus, when the condition charge packet is absent, the charge packet in the source potential well will flow through the corresponding gate potential well and then into the corresponding destination potential well. When the condition charge packet is present, then the corresponding gate potential will be increased and the charge packet in the source potential well will be prevented from moving into the corresponding destination potential well.

In this case the charge packet in the source potential well will find there is no barrier preventing it from moving into the other destination potential well. Some clock signals are marked with a prime, as F'. The clock lines marked with a prime are the same as the those that are not so marked except for their DC bias level. Because the condition shift registers and the source/destination shift registers share the metalization for the F180 clock, and because basically different things must occur when a charge packet passes beneath the F180 metalization in the two cases, the DC bias levels for the two sets of clock signals need to be separately adjustable.

Otherwise, when the gate electrode blocked a charge packet from moving from the source to the unselected destination, the same voltages applied to the metalizations on the corresponding condition shift register would likewise prevent the condition charge packet from moving along the condition shift register: this, of course, would cause problems. By adjusting the DC bias on the condition register clocks, we can insure that condition charge packets will move from left to right in a regular manner regardless of the voltage variation on the F180 clock caused by the presence or absence of the condition charge packet.

In the same way, by adjusting the bias on the source/destination clocks, we can insure that the voltage variation on the gate electrodes will prevent charge packets from moving into the wrong destination, thus insuring reliable switching action. In a real circuit, many RCT devices would be interconnected by a complex pattern of CCD shift registers. In most cases, this will force the data in one CCD shift register to cross over other shift registers. The use of a planar geometry makes the design of such a cross-over non-obvious.

A relatively simple approach would be to use two crossing sets of metalizations which define two CCD shift registers, and then apply a bias voltage to inactivate one of the shift registers.

A charge packet would be carried along the active shift register and would not be diverted into the inactive shift register because the bias voltage applied to the inactive shift register would eliminate the potential wells into which the charge packet might fall. By alternately inactivating one or the other shift register, charge packets could be carried across in first one and then the other direction.

This can be likened to a traffic crossing with a traffic light, where two cars approaching at right angles cannot simultaneously occupy the intersection. The traffic light is first green in one direction, allowing one car to cross; and is then green in the other direction, allowing the other car to cross.

To summarize this planar RCT: a charge packet moving along a CCD shift register will be moved into one of two "offspring" or "descendant" shift registers. Which descendant is selected will depend on which of two condition shift registers is occupied by a charge packet. This operation is logically universal: it can be used to implement a Fredkin gate and therefore any reversible logic circuit. Just as this planar RCT is conceptually similar to a 5-bucket RCT, so too a planar version similar to the 4-bucket RCT would also be feasible and would have the practical advantage that only a single condition would be required to operate it.

The planar 5-bucket RCT suggested here is intended to show that planar RCT's using simple sine waves for clocking are feasible. The design of an actual device might well differ significantly.

NANOMATERIALS FOR THE FUTURE

A conceptual high-energy-density physics aim approximately 2 millimetres in diameter for apply at Livermore's National Ignition adroitness. Deuterium gas is encapsulated in high-atomic-number doped effervesce and deuterium ice and encased in an external beryllium–copper shell. Less than 20 years ago, nanomaterials—substances on the scale of one-billionth of a meter—were a curiosity. Today, their diverse technical applications appear to be endless. Materials reduced to the nanoscale show identical a different active, chemical, and mechanical attributes compared to those they exhibit on a macroscale.

For example, nanoparticles have electronic structures that lie between that of bulk materials and atomic or molecular structures. Nanomaterials have real potential for impact in applications such as alternative energy conversion and energy harvesting.

The development of new nanomaterials will also lead to new mass-produced consumer products in cosmetics, cars, building components, clothes, health, and new smart electronic devices. Novel assembly and manipulation of nanostructured materials in confined geometries with tailored composition and function will have applications in catalysis, sensing, hydrogen storage, advanced nuclear materials, corrosion-resistant coatings, and photonics. The design of new nanomaterials with novel properties and functionalities requires atomic-level control of the material's structure.

The fabrication of functional nanomaterials can be achieved by conventional top-down methods such as lithography or bottom-up approaches such as self-assembly. The latter approach can be very cost-effective for the fabrication of macroscopic three-dimensional nanostructures. However, self-assembly is scientifically very challenging because it requires detailed understanding of materials interactions at the atomic level.

The goal of this LDRD project is twofold. The research team is advancing the scientific field of functional nanomaterials by combining the best of directed self-assembly with automated lithography and assembly to develop nanomaterials for fusion-energy applications for advanced laser systems. Second, they are applying this scientific knowledge to fabricate complex laser-fusion target structures by high-throughput techniques such as automated assembly and templating. Success will lay the scientific foundation for efficient production of laser fusion targets with unprecedented complexity. A successful experiment at fusion-class laser facilities such as the National Ignition Facility creates a unique laboratory environment previously only available in a nuclear test.

First among the unique characteristics is an inertially confined burning hydrogen plasma. Second is the bright neutron source that such a plasma will create. The challenge is to create laser targets that take advantage of this environment. To carry out experiments in this environment, researchers must place nanostructured materials with controlled composition and precise geometry in cubic millimetre spaces inside an inertial-confinement fusion capsule. Inertial fusion ignition will enable important laboratory and national missions, as well as new science. A series of such experiments is proposed for the National Ignition Facility, and NNSA's top scientific priority is significant advancement in the understanding of boost physics within the next five years.

Thus, among the first proposed experiments is a suite of boost physics experiments. These are designed to investigate the mix in burning plasmas. Complex targets for study of the mix in burning hydrogen plasmas, nuclear physics in high-neutron-brightness environments, and fast-ignition inertial-confinement fusion require precisely placing nanoporous materials with small amounts of additional incorporated elements to alter the material's properties (a process known as doping), inside the target capsule. Traditionally, target capsules are made by assembling prefabricated and machined components of the structure.

This approach limits the complexity of the structures that can be achieved. With self-assembly techniques, complex targets could be built either from the inside out or outside in. In the latter case, the ablator shell with a fill hole would serve as both a template and a container in which materials could be grown or modified.

LDRD researchers will use this "chemistry in a capsule" approach to generate complex nanostructures within the ablator shell by in situ self-assembly of functional building blocks. For example, a uniform and conforming film of a porous material lining the inside of the shell could be prepared by spin coating and subsequent curing by light, heat, or chemical initiation. Once formed, the properties of a nanoporous capsule liner could be further fine-tuned by atomic layer deposition. Possible applications include doping with mid-to-high-atomic-number elements as well as surface engineering of the wetting behaviour. In principle, this self-assembly technique could be applied to any nanostructure formation. In just the first year of this project, the LDRD team has made significant progress in all these areas.

They have successfully demonstrated the feasibility of the chemistry in a capsule approach by loading a diamond laser target with the nanomaterial precursor solution and forming nanoporous foam inside the capsule. Team members installed an atomic-layer-deposition system and demonstrated that the technique can be applied to deposit mid-to-high-atomic-number elements on the nanoporous foam liner inside the capsule. They also demonstrated that the capsule liner can be rendered inactive and then reactivated for atomic layer deposition.

This discovery opens the door to generating nanoporous foams with three-dimensional doping patterns by area-selective atomic layer deposition. The effects and requirements of this tailored doping on yield and symmetry of foam-filled capsules was assessed by hydrodynamic simulations.

X-ray scattering was used to study the effects of wetting organic aerogels with liquid hydrogen on the foam morphology. These experiments demonstrated that even low-density aerogels can survive liquid hydrogen wetting and de-wetting cycles. Finally, an in situ film-stress-monitoring capability was developed to study the microstructure-stress correlation in thick metal films grown by magnetron sputtering. The team was able to control the texture—and thus the stress of thick beryllium films—by applying a bias to the substrate, thereby changing the initial stress state from tensile to compressive.

Besides this impressive progress, they also developed novel nanomaterials for energy storage and conversion applications and studied their unique properties. Future LDRD efforts will include casting of thin films of aerogels in target capsules by growing cross-linkable polymer brushes, for example. Using in situ x-ray techniques, the team will study the effect of aerogel morphology on wetting with liquid hydrogen.

They will develop novel atomic-layer-deposition precursor chemistries for doping with high-atomic-number materials. Area-selective doping will be attempted through spatially selective modifications of the substrate's surface chemistry. In addition, new functionalities will be added to their nanomanipulator to measure thickness profiles from three-dimensional targets and to apply picoliter amounts of liquids. Researchers will continue work on stress control in metal deposition and develop predictive multiscale tools. The successful conclusion of this LDRD project will enable the fabrication of complex targets for experiments to understand boost physics, investigate the dynamics of nuclear excited states, and pursue inertial-confinement fusion fast ignition.

It will also create new materials for catalysis, hydrogen storage, and self-healing nuclear reactor materials. A very important long-term benefit of this effort will be the establishment of an outstanding Laboratory team for the design and assembly of tailored nanomaterials and nanostructures.

7

Toxicology of Nanoparticles

BACKGROUND

Nanotoxicology is a sub-specialty of particle toxicology. It addresses the toxicology of nanoparticles which appear to have toxicity effects that are unusual and not seen with larger particles. Nanoparticles can be divided into combustion-derived nanoparticles, manufactured nanoparticles like carbon nanotubes and naturally occurring nanoparticles from volcanic eruptions, atmospheric chemistry etc. Typical nanoparticles that have been studied are titanium dioxide, alumina, zinc oxide, carbon black, and carbon nanotubes, and "nano-C_{60}". Nanoparticles have much larger surface area to unit mass ratios which in some cases may lead to greater pro-inflammatory effects. In addition, some nanoparticles seem to be able to translocate from their site of deposition to distant sites such as the blood and the brain. This has resulted in a sea-change in how particle toxicology is viewed-instead of being confined to the lungs, nanoparticle toxicologists study the brain, blood, liver, skin and gut. Nanotoxicology has revolutionised particle toxicology and rejuvenated it.

REACTIVE OXYGEN SPECIES

For some types of particles, the smaller they are, the greater their surface area to volume ratio and the higher their chemical reactivity and biological activity. The greater chemical reactivity of nanomaterials can result in increased production of *reactive oxygen species* (ROS), including free radicals. ROS production has been found in a diverse range of nanomaterials including carbon fullerenes, carbon nanotubes and nanoparticle metal oxides. ROS and free radical production is one of the primary mechanisms of nanoparticle toxicity; it may result in oxidative stress, inflammation, and consequent damage to proteins, membranes and DNA.

Biodistribution

The extremely small size of nanomaterials also means that they much more readily gain entry into the human body than larger sized particles. How these nanoparticles behave inside the body is still a major question that needs to be resolved. The behaviour of nanoparticles is a function of their size, shape and surface reactivity with the surrounding tissue. In principle, a large number of particles could overload the body's phagocytes, cells that ingest and destroy foreign matter, thereby triggering stress reactions that lead to inflammation and weaken the body's defence against other pathogens.

In addition to questions about what happens if non-degradable or slowly degradable nanoparticles accumulate in bodily organs, another concern is their potential interaction or interference with biological processes inside the body. Because of their large surface area, nanoparticles will, on exposure to tissue and fluids, immediately adsorb onto their surface some of the macromolecules they encounter. This may, for instance, affect the regulatory mechanisms of enzymes and other proteins. Nanomaterials are able to cross biological membranes and access cells, tissues and organs that larger-sized particles normally cannot.

Nanomaterials can gain access to the blood stream via inhalation or ingestion. At least some nanomaterials can penetrate the skin; even larger microparticles may penetrate skin when it is flexed. Broken skin is an ineffective particle barrier, suggesting that acne, eczema, shaving wounds or severe sunburn may accelerate skin uptake of nanomaterials. Then, once in the blood stream, nanomaterials can be transported around the body and be taken up by organs and tissues, including the brain, heart, liver, kidneys, spleen, bone marrow and nervous system. Nanomaterials have proved toxic to human tissue and cell cultures, resulting in increased oxidative stress, inflammatory cytokine production and cell death. Unlike larger particles, nanomaterials may be taken up by cell mitochondria and the cell nucleus. Studies demonstrate the potential for nanomaterials to cause DNA mutation and induce major structural damage to mitochondria, even resulting in cell death.

Nanotoxicity Studies

Since there is no authority to regulate nanotech-based products, there are many products that could possibly be dangerous to humans. Scientific research has indicated the potential for some nanomaterials to be toxic to humans or the environment. In March 2004 tests conducted by environmental toxicologist Eva Oberdörster, Ph.D. working with Southern Methodist University in Texas, found extensive brain damage to fish exposed to fullerenes for a period of just 48 hours at a relatively moderate dose of 0.5 parts per million. The fish also exhibited changed gene markers in their livers, indicating their entire physiology was affected.

In a concurrent test, the fullerenes killed water fleas, an important link in the marine food chain. The extremely small size of fabricated nanomaterials also means that they are much more readily taken up by living tissue than presently known toxins. Nanoparticles can be inhaled, swallowed, absorbed through skin and deliberately or accidentally injected during medical procedures. They might be accidentally or inadvertently released from materials implanted into living tissue.

Researcher Shosaku Kashiwada of the National Institute for Environ-mental Studies in Tsukuba, Japan, in a more recent study, intended to further investigate the effects of nanoparticles on soft-bodied organisms. His study allowed him to explore the distribution of water-suspended fluorescent nanoparticles throughout the eggs and adult bodies of a species of fish, known as the see-through medaka. Moreover, small fish like the see-through medaka have been popular test subjects for human diseases and organogenesis for other reasons as well, including their transparent embryos, rapid embryo development, and the functional equivalence of their organs and tissue material to that of mammals.

Because the see-through medaka have transparent bodies, analysing the deposition of fluorescent nanoparticles throughout the body is quite simple. For his study, Dr. Kashiwada evaluated four aspects of nanoparticle accumulation. These included the overall accumulation and the size-dependent accumulation of nanoparticles by medaka eggs, the effects of salinity on the aggregation of nanoparticles in solution and on their accumulation by medaka eggs, and the distribution of nanoparticles in the blood and organs of adult medaka. It was also noted that nanoparticles were in fact taken up into the bloodstream and deposited throughout the body. In the medaka eggs, there was a high accumulation of nanoparticles in the yolk; most often bioavailibility was dependent on specific sizes of the particles.

Adult samples of medaka had accumulated nanoparticles in the gills, intestine, brain, testis, liver, and bloodstream. One major result from this study was the fact that salinity may have a large influence on the bioavailibility and toxicity of nanoparticles to penetrate membranes and eventually kill the specimen. As the use of nanomaterials increases worldwide, concerns for worker and user safety are mounting. To address such concerns, the Swedish Karolinska Institute conducted a study in which various nanoparticles were introduced to human lung epithelial cells. The results, released in 2008, showed that iron oxide nanoparticles caused little DNA damage and were non-toxic. Zinc oxide nanoparticles were slightly worse. Titanium dioxide caused only DNA damage. Carbon nanotubes caused DNA damage at low levels. Copper oxide was found to be the worst offender, and was the only nanomaterial identified by the researchers as a clear health risk.

No Fullerene Toxicity Reported

Nanoparticles can also be made of C_{60}, as is the case with almost any room temperature solid, and several groups have done this and studied toxicity of such particles. The results in the work of Oberdörster at Southern Methodist University, published in "Environmental Health Perspectives" in July 2004, in which questions were raised of potential cytotoxicity, has now been shown by several sources to be likely caused by the tetrahydrofuran used in preparing the 30 nm–100 nm particles of C_{60} used in the research. Isakovic, *et al.*, 2006, who review this phenomenon, gives results showing that removal of THF from the C_{60} particles resulted in a loss of toxicity.

Sayes, *et al.*, 2007, also show that particles prepared as in Oberdorster caused no detectable inflammatory response when instilled intratracheally in rats after observation for 3 months, suggesting that even the particles prepared by Oberdorster do not exhibit markers of toxicity in mammalian models. This work used as a benchmark quartz particles, which did give an inflammatory response. A comprehensive and recent review of work on fullerene toxicity is available in "Toxicity Studies of Fullerenes and Derivatives," a chapter from the book "Bio-applications of Nanoparticles". In this work, the authors review the work on fullerene toxicity beginning in the early 1990s to present, and conclude that the evidence gathered since the discovery of fullerenes overwhelmingly points to C_{60} being non-toxic. As is the case for toxicity profile with any chemical modification of a structural moiety, the authors suggest that individual molecules be assessed individually.

Immunogenicity of Nanoparticles

Very little attention has been directed towards the potential immunogenicity of nanostructures. Nanostructures can activate the immune system inducing inflammation, immune responses, allergy, or even affect to the immune cells in a deleterious or beneficial way. More studies are needed in order to know the potential deleterious or beneficial effects of nano-structures in the immune system. In comparison to conventional pharmeceutical agents, nanostructures have very large sizes and immune cells, especially phagocytic cells, recognize and try to destroy them.

Complications with Nanotoxicity Studies

Size is therefore a key factor in determining the potential toxicity of a particle. However it is not the only important factor. Other properties of nanomaterials that influence toxicity include: chemical composition, shape, surface structure, surface charge, aggregation and solubility, and the presence or absence of functional groups of other chemicals. The large number of variables influencing toxicity means that it is difficult to generalise about health risks associated with exposure to nanomaterials–each new nanomaterial must be assessed individually and all material properties must be taken into account.

In addition, standarization of toxicology tests between laboratories are needed. Díaz, B. *et al.* from the University of Vigo has shown that many different cell lines should be studied in order to know if a nanostructure induces toxicity, and human cells can internalize aggregated nanoparticles. Moreover, it is important to take into account that many nanostructures aggregate in biological fluids, but groups manufacturing nanostructures do not care much about this matter. Many efforts of interdisciplinary groups are strongly needed in order to progress in this field.

Effect of Aggregation/Agglomeration of Nanoparticles

Many nanoparticles agglomerate or aggregate when they are placed in environmental or biological fluids. The terms agglomeration and aggregation have distinct definitions according to the standards organizations ISO and ASTM, where agglomeration signifies more loosely bound particles and aggregation signifies very tightly bound or fused particles. Nanoparticles frequently agglomerate due to the high ionic strength of environmental and biological fluids, which shields the repulsion due to charges on the nanoparticles. Unfortunately, agglomeration has frequently been ignored in nanotoxicity studies, even though agglomeration would be expected to affect nanotoxicity since it changes the size, surface area, and sedimentation properties of the nanoparticles.

In addition, many nanoparticles will agglomerate to some extent in the environment or in the body before they reach their target, so it is desirable to study how toxicity is affected by agglomeration. A method was published that can be used to produce different mean sizes of stable agglomerates of several metal, metal oxide, and polymer nanoparticles in cell culture media for cell toxicity studies. Different mean sizes of agglomerates are produced by allowing the nanoparticles to agglomerate to a particular size in cell culture media without protein, and then adding protein to coat the agglomerates and "freeze" them at that size. By waiting different amounts of time before adding protein, different mean sizes of agglomerates of a single type of nanoparticle can be produced in an otherwise identical solution, allowing one to study how agglomerate size affects toxicity. In addition, it was found that vortexing while adding a high concentration of nanoparticles to the cell culture media produces much less agglomerated nanoparticles than if the dispersed solution is only mixed after adding the nanoparticles.

Challenges of the Nano-Visualisation and Related Unknowns in Nanotoxicology

With comparison to more conventional toxicology studies, the nano-toxicology field is however suffering form a lack of easy characterisation of the potential contaminants, the "nano" scale been still a scale difficult to apprehend. The biological systems are themselves still not completely known at this scale. Ultimate Atomic visualisation methods such as Electron microscopy and *Atomic force Microscopy* (AFM) analysis are allowing fantastic progresses in the visualisation of the nano world.

Yet, further nanotoxicology studies will require extremely precise characterisation of the specificities of a given nano-element: size, chemical composition, detailed shape, level of aggregation, combination with other vectors, etc. These properties would have to be determined not only on the nanocomponent before its introduction in the living environnment but also in the biological environnement. This is why nanotoxicoly is a fantastic field of research. This is also why it is not easy to determine to what extent a given nanoparticule has a dramatic effect when compared to comparable nanoparticules already present in our environnement either through natural/ biological origin or through ancestral human activity.

NANOPARTICLES

A nanoparticle (or nanopowder or nanocluster or nanocrystal) is a microscopic particle with at least one dimension less than 100 nm.There is no accepted international definition of a nanoparticle, but one given in the new PAS71 document developed in the UK is: "A particle having one or more dimensions of the order of 100nm or less".

There is a note associated with this definition: "Novel properties that differentiatenanoparticles from the bulk material typically develop at a critical length scale of under 100nm".

The "novel properties" mentioned are entirely dependent on the fact that at the nano-scale, the physics of nanoparticles mean that their properties are different from the properties of the bulk material. This makes the size of particles or the scale of its features the most important attribute of nanoparticles.

There is no strict dividing line between nanoparticles and non-nanoparticles. The size at which materials display different properties to the bulk material is material dependant and can certainly be claimed for many materials much larger in size than 100nm. Definitions certainly become more difficult for materials that are a very long way from being a sphere, such as carbon nanotubes for example. One of the aims for these materials is to grow them into long tubes, certainly not 'nano' in length, but as they have a diametre in the order of 3nm for a single walled tube, they have properties that distinguish them from other allotropes of carbon, and hence can be described as 'nanomaterials'. This sort of nanomaterial has led to the extension of the idea of nanomaterials being considered as such if any one of their structural features are on a scale of less than 100nm, that cause their properties to be different from that of the bulk material.

Nanoparticle research is currently an area of intense scientific research, due to a wide variety of potential applications in biomedical, optical, and electronic fields. Nanoparticles are of great scientific interest as they are effectively a bridge between bulk materials and atomic or molecular structures. A bulk material should have constant physical properties regardless of its size, but at the nano-scale this is often not the case.

Size-dependent properties are observed such as quantum confinement in semiconductor particles, surface plasmon resonance in some metal particles and superparamagnetism in magnetic materials. The properties of materials change as their size approaches the nanoscale and as the percentage of atoms at the surface of a material becomes significant.

For bulk materials larger than one micrometre the percentage of atoms at the surface is minuscule relative to the total number of atoms of the material. The interesting and sometimes unexpected properties of nanoparticles are not partly due to the aspects of the surface of the material dominating the properties in lieu of the bulk properties. Nanoparticles exhibit a number of special properties relative to bulk material. For example, the bending of bulk copper (wire, ribbon, etc.) occurs with movement of copper atoms/clusters at about the 50 nm scale. Copper nanoparticles smaller than 50 nm are considered super hard materials that do not exhibit the same malleability and ductility as bulk copper.

The change in properties is not always desirable. Ferroelectric materials smaller than 10 nm can switch their magnetisation direction using room temperature thermal energy, thus making them useless for memory storage. Suspensions of nanoparticles are possible because the interaction of the particle surface with the solvent is strong enough to overcome differences in density, which usually result in a material either sinking or floating in a liquid.

Nanoparticles often have unexpected visible properties because they are small enough to confine their electrons and produce quantum effects. For example gold nanoparticles appear deep red to black in solution. Nanoparticles have a very high surface area to volume ratio.

This provides a tremendous driving force for diffusion, especially at elevated temperatures. Sintering can take place at lower temperatures, over shorter time scales than for larger particles. This theoretically does not affect the density of the final product, though flow difficulties and the tendency of nanoparticles to agglomerate complicates matters.

The large surface area to volume ratio also reduces the incipient melting temperature of nanoparticles. Adding just the right dash of nanoparticles to standard mixes of lubricants and refrigerants could yield the equivalent of an energy-saving chill pill for factories, hospitals, ships, and others with large cooling systems, suggest the latest results from National Institute of Standards and Technology (NIST) research that is pursuing promising formulations. NIST experiments with varying concentrations of nanoparticle additives indicate a major opportunity to improve the energy efficiency of large industrial, commercial, and institutional cooling systems known as chillers.

These systems account for about 13 per cent of the power consumed by the nation's buildings, and about 9 per cent of the overall demand for electric power, according to the Department of Energy. NIST researcher Mark Kedzierski has found that dispersing "sufficient" amounts of copper oxide particles (30 nanometres in diametre) in a common polyester lubricant and combining it with an equally pedestrian refrigerant (R134a) improves heat transfer by between 50 per cent and 275 per cent. "We were astounded," he says.

Results of this work have been presented at recent conferences and will be reported in an upcoming issue of the ASME *Journal of Heat Transfer*. Just how nanomaterial additives to lubricants improve the dynamics of heat transfer in refrigerant/lubricant mixtures is not thoroughly understood. The NIST research effort aims to fill gaps in knowledge that impede efforts to determine and, ultimately, predict optimal combinations of the three types of substances.

"As with all good things, the process is far from foolproof," Kedzierski explains. "In fact, an insufficient amount or the wrong type of particles might lead to degradation in performance."

On the basis of work so far, the researcher speculates several factors likely account for nanoparticle-enabled improvements in heat-transfer performance. For one, nanoparticles of materials with high thermal conductivity improve heat transfer rates for the system. Preliminary results of the NIST research also indicate that, in sufficient concentrations, nanomaterials enhance heat transfer by encouraging more vigourous boiling of the mixture.

The tiny particles stimulate, in effect, double bubbles—secondary bubbles that form atop bubbles initiated at the boiling site. Bubbles carry heat away from the surface, and the fact that they're being formed more efficiently because of the nanoparticles means the heat gets transferred more readily.

Other interactions, Kedzierski says, also are likely to contribute to the dramatic performance improvements reported at NIST and elsewhere. Success in optimizing recipes of refrigerants, lubricants and nanoparticle additives would pay immediate and long-term dividends. If they did not harm other aspects of equipment performance, high-performance mixtures could be swapped into existing chillers, resulting in immediate energy savings. And, because of improved energy efficiency, next-generation equipment would be smaller, requiring fewer raw materials in their manufacture. Researchers at the National Institute of Standards and Technology (NIST) and three collaborating institutions are using a new laboratory model of the membrane surrounding neurons in the brain to study how a protein long suspected of a role in early-stage Alzheimer's disease actually impairs a neuron's structure and function.

The team's findings are reported in a new paper in the *Biophysical Journal*. The brain's neurons transmit nerve impulses down a long stem that is surrounded by a two-layer membrane. In the neuron's normal, "rest" state, this membrane actively sorts sodium ions to the outside of the cell and potassium ions to the inside. To transmit a nerve impulse, an electrochemical change ripples down the membrane in advance of the impulse, making it temporarily more permeable and allowing the ions to swap places.

That in turn changes the electrical potential across the membrane, allowing the impulse to pass. Afterwards, the membrane returns to rest and begins sorting the ions again. Medical experts have hypothesized for years that small polypeptides called amyloid beta peptides somehow create a "leaky" membrane that disrupts this balanced back-and-forth switching of the electrical potential and, in turn, normal impulse transmission. Alzheimer's disease—the progressive brain disorder that is the nation's sixth leading cause of death—is believed to start with such breakdowns. As the disease progresses, amyloid beta peptides clump together to form plaques that further destroy nerve function.

Studying the beginnings of Alzheimer's is nearly impossible in humans because by the time the disease is diagnosed, most patients have moved into its later stages. Researchers at NIST have developed a laboratory model that recreates a simplified version of the nerve cell membrane, allowing the study of Alzheimer's disease mechanisms at the molecular level. A clever piece of molecular-level design, the system is built by first covering a silica surface with gold. Sulfur atoms, which bond well to gold, are then added to act as anchors to hold the bilayer membrane. The result is a stable, tethered membrane with an aqueous environment on both sides that accurately models the behaviour of the nerve cell membrane.

A collaborative team of researchers from NIST, Carnegie Mellon University, the University of California-Irvine and the Biochemistry Institute (BCHI) in Vilnius, Lithuania, exposed the membrane model to different concentrations of a specific form of amyloid beta peptides comprised of soluble, tiny (5-6 nanometres, approximately twice the diametre of a DNA helix) chains. The researchers found increased cation movement across the normally strong barrier at the higher concentrations of the peptides. The data support the hypothesis that membrane "leakiness" is not due to a permanent hole being formed but rather to an aggregation of amyloid beta peptides in the membrane that allows cations to be passed from peptide to peptide across the bilayer, like a baton handed off by relay runners.

The researchers are continuing to use their model system to better understand the role amyloid beta peptides play in early-stage Alzheimer's disease. Future plans include investigating how amyloid beta peptide aggregates arrange themselves in the membrane, how the peptide aggregates affect or influence calcium channels in the membrane, and how the peptides interact with membranes constructed with other types of lipids.

Titanium, a protean element with applications from pigments to aerospace alloys, could get a new role as an environmentally friendly additive for automotive oil, thanks to work by materials scientists from Afton Chemical Corporation and the National Institute of Standards and Technology (NIST). In a recent paper,* the researchers established that a titanium compound added to engine oil creates a wear-resistant nanoscale layer bound to the surface of vulnerable engine parts, making it a credible substitute for older compounds that do not coexist well with antipollution equipment.

Modern engine lubricating oil is a complex, highly engineered mixture, up to 20 per cent of which may be special additives to enhance properties such as viscosity and stability and to reduce sludge formation and engine wear, according to Afton specialists. For years antiwear additives for high-performance oils have been phosphorous compounds, particularly ZDDP,** that work by forming a polyphosphate film on engine parts that reduces wear. Unfortunately phosphorus is a chemical poison for automobile catalytic converters, reducing their effectiveness and life span, so industry chemists have been searching for ways to replace or reduce the use of ZDDP. It's not a simple problem because the additive has several useful functions in addition to wear resistance.

Titanium is one candidate replacement. Mechanical tests of an organic titanium compound at Afton demonstrated that it provided superior wear resistance when added to a fully formulated engine oil, suggesting that oil chemists could use less ZDDP. Just how the titanium compound works was an open question, however. Surface analysis tests could detect titanium in the wear tracks of test surfaces but not with enough sensitivity to determine its chemical nature—and whether, for example, it was just lying there or bound to the metal surface. To resolve the issue, the researchers turned to NIST's soft X-ray beamline at the National Synchrotron Light Source (NSLS) in Brookhaven, N.Y.

The NIST beamline instruments use low-energy ("soft") X-rays that can be precisely tuned to specific elements to measure chemical bonds both at the surface of a sample and deeper into the bulk of the material. Powered by the NSLS, the facility is at least 10 times more sensitive than commonly available instruments. The measurements revealed that the antiwear enhancement comes from titanium chemically bound into the metal structure of the engine surface, forming a hard oxide, iron titanate. Comparing the test data to that of several possible compounds, the research team was able to identify the specific oxide. While considerably more work remains to be done, the results suggest that titanium could play an important role in future low-phosphorus lubricating oils.

Scientists at the National Institute of Standards and Technology (NIST) have developed a technique to measure the formation of clumps of proteins in protein-based pharmaceuticals. This first systematic study* clarifies the conditions under which scientists can be assured that their instruments are faithfully measuring the formation of protein aggregates, a major concern because of its impact on quality control and safety in biologic drug manufacturing.

Proteins, a main constituent of many new drugs, are large molecules that have a tendency to stick to each other and form clumps during their manufacture. These clumps have been associated with severe immune responses. In at least one case, the inadvertent creation of protein clumps during the processing of a drug to treat anemia caused an immune reaction in about 250 patients that destroyed their ability to produce red blood cells. Those patients now have to receive blood transfusions every few days to replenish these vital cells.

Events such as this led the Food and Drug Administration to call for the development of sensitive and rapid measurement tools that can detect aggregation of protein drugs during the manufacturing phase. To address the problem, NIST researchers hit upon the idea of adapting a technique known as electrospray differential mobility analysis (ES-DMA). Commonly used to size soot and other aerosols, ES-DMA uses an electric current to vapourize a solution of proteins into tiny charged water droplets, each containing a single protein molecule or protein aggregate. Once these droplets evapourate, the charged proteins and protein aggregates are drawn into an oppositely charged tube. By making controlled adjustments to the voltage of the tube and the velocity of the air flowing through it, researchers can collect particles of a specific size, allowing the proteins and protein aggregates to be precisely sorted and counted.

The NIST team adapted the technique for biopharmaceutical applications. According to researcher Leonard Pease, ES-DMA is tricky to get right, but the NIST team was able to define the conditions needed to electrospray proteins and protein aggregates reliably and repeatedly. NIST scientists favour ES-DMA for its ability to quickly resolve particle sizes differing by as little as 0.2 nanometres, to provide a direct measure of particle size distributions, and to accept bioreactor samples with significantly reduced preparation requirements.

In addition to sizing proteins, Pease said the technique could also be used to accurately size many different types of particles used in medicine, such as the viruses used in the human papillomavirus (HPV) vaccine and gene therapy. "The adaptation of this technology is just one of many excellent examples of how NIST seeks out and works with U.S. industry," says Willie E. May, director of the NIST Chemical Science and Technology Laboratory.

Data breaches are a recurring nightmare for IT managers responsible for securing not only their company's confidential data, but possibly also sensitive information belonging to their clients, such as social security numbers or health or financial records. To help managers safeguard valuable information most efficiently, computer scientists at the National Institute of Standards and Technology (NIST) are applying security metrics to computer network pathways to assign a probable risk of attack to guide IT managers in securing their networks.

"We analyse all of the paths that system attackers could penetrate through a network," says computer scientist Anoop Singhal, "and assign a risk to each component of the system. Decision makers can use our assigned probabilities to make wise decisions and investments to safeguard their network." The research was presented at a conference earlier this month.*

Computer networks are made up of components varying from individual computers, to servers and routers. Once inside a network's firewall, for a seemingly mild-mannered purpose as posting an image to a file transfer protocol (FTP) site, a hacker can travel through the network through a variety of routes to hit the jackpot of valuable data. In addition to hardware, the hacker can break in through software on the computers, especially file-sharing applications that have been blamed for some major data breaches recently. NIST researchers evaluate each route and assign it a risk based on how challenging it is to the hacker. The paths are determined using a technique called "attack graphs." A new analysis technique based on attack graphs was jointly developed by Singhal and research colleagues at George Mason University. A patent is pending on the technique.

Singhal and his team determine risk by using these attack graphs and NIST's National Vulnerability Database (NVD). This government repository includes a collection of security-related software weaknesses that hackers can exploit. NVD data was collected from software vendors and scores are assigned from most to least insecure by experts.

For example in a simple system there is an attacker on a computer, a firewall, router, an FTP server and a database server. The goal for the attacker is to find the simplest path into the jackpot—the database server. Attack Graph Analysis determines three potential attack paths. For each path in the graph, the NIST researchers assign an attack probability based on the score in the NVD database.

Because it takes multiple steps to reach the goal, the probabilities of each component are multiplied to determine the overall risk. One path takes only three steps. The first step has an 80 per cent chance of being hacked, the second, a 90 per cent chance. The final step requires great expertise, so there is only a 10 per cent probability it can be breached. By multiplying the three probabilities together, that path is pretty secure with a less than 10 per cent chance of being hacked.

CHEMICAL PROCESSING AND SYNTHESIS

The chemical processing and synthesis of high performance technological components for the private, industrial, and military sectors requires the use of high purity ceramics, polymers, glass-ceramics and material composites. In condensed bodies formed from fine powders, the irregular particle sizes and shapes in a typical powder often lead to non-uniform packing morphologies that result in packing density variations in the powder compact.

Uncontrolled agglomeration of powders due to attractive van der Waals forces can also give rise to in microstructural inhomogeneities. Differential stresses that develop as a result of non-uniform drying shrinkage are directly related to the rate at which the solvent can be removed, and thus highly dependent upon the distribution of porosity. Such stresses have been associated with a plastic-to-brittle transition in consolidated bodies, and can yield to crack propagation in the unfired body if not relieved.

In addition, any fluctuations in packing density in the compact as it is prepared for the kiln are often amplified during the sintering process, yielding inhomogeneous densification. Some pores and other structural defects associated with density variations have been shown to play a detrimental role in the sintering process by growing and thus limiting end-point densities. Differential stresses arising from inhomogeneous densification have also been shown to result in the propagation of internal cracks, thus becoming the strength-controlling flaws.

Inert gas evaporation and inert gas deposition are free many of these defects due to the distillation (c.f. purification) nature of the process and having enough time to form single crystal particles, however even their non-aggreated deposits have lognormal size distribution, which is typical with nanoparticles. The reason why modern gas evaporation techniques can produce a relatively narrow size distribution is that aggregation can be avoided. However, even in this case, random residence times in the growth zone, due to the combination of drift and diffusion, result in a size distribution appearing lognormal.

It would therefore appear desirable to process a material in such a way that it is physically uniform with regard to the distribution of components and porosity, rather than using particle size distributions which will maximize the green density. The containment of a uniformly dispersed assembly of strongly interacting particles in suspension requires total control over interparticle forces. Monodisperse nanoparticles and colloids provide this potential.

Monodisperse powders of colloidal silica, for example, may therefore be stabilized sufficiently to ensure a high degree of order in the colloidal crystal or polycrystalline colloidal solid which results from aggregation. The degree of order appears to be limited by the time and space allowed for longer-range correlations to be established. Such defective polycrystalline colloidal structures would appear to be the basic elements of submicrometer colloidal materials science, and, therefore, provide the first step in developing a more rigorous understanding of the mechanisms involved in microstructural evolution in high performance materials and components.

SYNTHESIS OF CARBON NANOTUBES

A range of methodologies have been developed to produce nanotubes in sizeable quantities, including arc discharge, laser ablation, high pressure carbon monoxide (HiPco), and vapour liquid solid (VLS) growth. All these processes take place in vacuum or at low pressure with a process gases, although VLS growth can take place at atmospheric pressure.

Large quantities of nanotubes can be synthesized by these methods; advances in catalysis and continuous growth processes are making SWNTs more commercially viable. The first observation of nanotubes was in the carbon soot formed during the arc discharge production of fullerenes. The high temperatures caused by the discharge caused the carbon contained in the negative electrode to sublime and the CNTs are deposited on the opposing electrode. Tubes produced by this method were initially multi walled tubes (MWNTs). However, with the addition of cobalt to the vapourized carbon, it is possible to grow single walled nanotubes.

This method it produces a mixture of components, and requires further purification to separate the CNTs from the soot and the residual catalytic metals. Producing CNTs in high yield depends on the uniformity of the plasma arc, and the temperature of the deposit forming on the carbon electrode. Higher yield and purity of SWNTs may be prepared by the use of a dual-pulsed laser. SWNTs can be grown in a 50 per cent yield through direct vapourization of a Co/Ni doped graphite rod with a high-powered laser in a tube furnace operating at 1200 °C.

The material produced by this method appears as a mat of "ropes", 10 - 20 nm in diametre and up to 100 ìm or more in length. Each rope consists of a bundle of SWNTs, aligned along a common axis. By varying the process parametres such as catalyst composition and the growth temperature, the average nanotube diametre and size distribution can be varied. Although arc-discharge and laser vapourization are currently the principal methods for obtaining small quantities of high quality SWNTs, both methods suffer from drawbacks.

The first is that they involve evapourating the carbon source, making scale-up on an industrial level difficult and energetically expensive. The second issue relates to the fact that vapourization methods grow SWNTs in highly tangled forms, mixed with unwanted forms of carbon and/or metal species. The SWNTs thus produced are difficult to purify, manipulate, and assemble for building nanotube-device architectures for practical applications.

In order to overcome some of the difficulties of these high-energy processes, the chemical catalysis method was developed in which a hydrocarbon feedstock is used in combination with a metal catalyst. The catalyst is typically, but not limited to iron, colbalt, or iron/molybdenum, it is heated under reducing conditions in the presence of a suitable carbon feedstock, *e.g.*, ethylene.

This method can be used for both SWNTs and MWNTs; the formation of each is controlled by the identity of the catalyst and the reaction conditions. A convenient laboratory scale apparatus is available from Nanotech Innovations, Inc., for the synthesis of highly uniform, consistent, research sample that uses pre-weighed catalyst/carbon source ampoules.

This system, allows for 200 mg samples of MWNTs to be prepared for research and testing. The use of CO as a feedstock, in place of a hydrocarbon, led to the development of the high-pressure carbon monoxide (HiPco) procedure for SWNT synthesis. By this method, it is possible to produce gram quantities of SWNTs, unfortunately, efforts to scale beyond that have not met with complete success.

Initially developed for small-scale investigations of catalyst activity, vapour liquid solid (VLS) growth of nanotubes has been highly studied, and now shows promise for large-scale production of nanotubes. Recent approaches have involved the use of well-defined nanoparticle or molecular precursors and many different transition metals have been employed, but iron, nickel, and cobalt remain to be the focus of most research.

The nanotubes grow at the sites of the metal catalyst; the carbon-containing gas is broken apart at the surface of the catalyst particle, and the carbon is transported to the edges of the particle, where it forms the nanotube. The length of the tube grown in surface supported catalyst VLS systems appears to be dependent on the orientation of the growing tube with the surface.

By properly adjusting the surface concentration and aggregation of the catalyst particles it is possible to synthesize vertically aligned carbon nanotubes, *i.e.*, as a carpet perpendicular to the substrate. Of the various means for nanotube synthesis, the chemical processes show the greatest promise for industrial scale deposition in terms of its price/unit ratio.

There are additional advantages to the VLS growth, which unlike the other methods is capable of growing nanotubes directly on a desired substrate. The growth sites are controllable by careful deposition of the catalyst. Additionally, no other growth methods have been developed to produce vertically aligned SWNTs.

CHEMICAL FUNCTIONALIZATION OF CARBON NANOTUBES

The limitation of using carbon nanotubes in any practical applications has been its solubility; for example SWNTs have little to no solubility in most solvent due to the aggregation of the tubes. Aggregation/roping of nanotubes occurs as a result of the high van der Waals binding energy of *ca*. 500 eV per mm of tube contact. The van der Waals force between the tubes is so great, that it take tremendous energy to pry them apart, making it very to make combination of nanotubes with other materials such as in composite applications.

The functionalization of nanotubes, *i.e.*, the attachment of "chemical functional groups" provides the path to overcome these barriers. Functionalization can improve solubility as well as processibility, and has been used to align the properties of nanotubes to those of other materials. The clearest example of this is the ability to solubilize nanotubes in a variety of solvents, including water. It is important when discussing functionalization that a distinction is made between covalent and non-covalent functionalization.

Current methods for solubilizing nanotubes without covalent functionalization include highly aromatic solvents, super acids, polymers, or surfactants. Non-covalent "functionalization" is generally on the concept of supramolecular interactions between the SWNT and some macromolecule as a result of various adsorption forces, such as van der Waals' and n-stacking interactions. The chemical speciation of the nanotube itself is not altered as a result of the interaction. In contrast, covalent functionalization relies on the chemical reaction at either the sidewall or end of the SWNT.

As may be expected the high aspect ratio of nanotubes means that sidewall functionalization is much more important than the functionalization of the cap. Direct covalent sidewall functionalization is associated with a change of hybridization from sp^2 to sp^2 and a simultaneous loss of conjugation. An alternative approach to covalent functionalization involves the reaction of defects present (or generated) in the structure of the nanotube.

Defect sites can be the open ends and holes in the sidewalls, and pentagon and heptagon irregularities in the hexagon graphene framework (often associated with bends in the tubes). All these functionalizations are exohedral derivatizations. Taking the hollow structure of nanotubes into consideration, endohedral functionalization of SWNTs is possible, *i.e.*, the filling of the tubes with atoms or small molecules. It is important to note that covalent functionalization methods have one problem in common: extensive covalent functionalization modifies SWNT properties by disrupting the continuous n-system of SWNTs

Various applications of nanotubes require different, specific modification to achieve desirable physical and chemical properties of nanotubes. In this regard, covalent functionalization provides a higher degree of fine-tuning the chemistry and physics of SWNTs than non-covalent functionalization. Until now, a variety of methods have been used to achieve the functionalization of nanotubes.

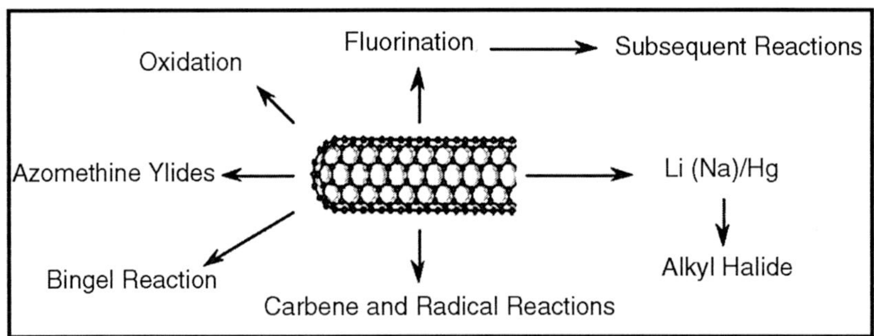

Fig. Schematic Description of Various Covalent Functionalization Strategies for SWNTs.

Taking chemistry developed for C_{60}, SWNTs may be functionalized using 1,3 dipolar addition of azomethine ylides. The functionalized SWNTs are soluble in most common organic solvents. The azomethine ylide functionalization method was also used for the purification of SWNTs. Under electrochemical conditions, aryl diazonium salts react with SWNTs to achieve functionalized SWNTs, alternatively the diazonium ions may be generated *in-situ* from the corresponding aniline, while a solvent free reaction provides the best chance for large-scale functionalization this way. In each of these methods it is possible to control the amount of functionalization on the tube by varying reaction times and the reagents used; functionalization as high as 1 group per every 10 - 25 carbon atoms is possible.

Organic functionalization through the use of alkyl halides, a radical pathway, on tubes treated with lithium in liquid ammonia offers a simple and flexible route to a range of functional groups. In this reaction, functionalization occurs on every 17 carbons. Most success has been found when the tubes are dodecylated. These tubes are soluble in chloroform, DMF, and THF.

The addition of oxygen moieties to SWNT sidewalls can be achieved by treatment with acid or wet air oxidation, and ozonolysis. The direct epoxidation of SWNTs may be accomplished by the direct reaction with a peroxide reagent, or catalytically. Catalytic de-epoxidation allows for the quantitative analysis of sidewall epoxide and led to the surprising result that previously assumed "pure" SWNTs actually contain *ca*. 1 oxygen per 250 carbon atoms.

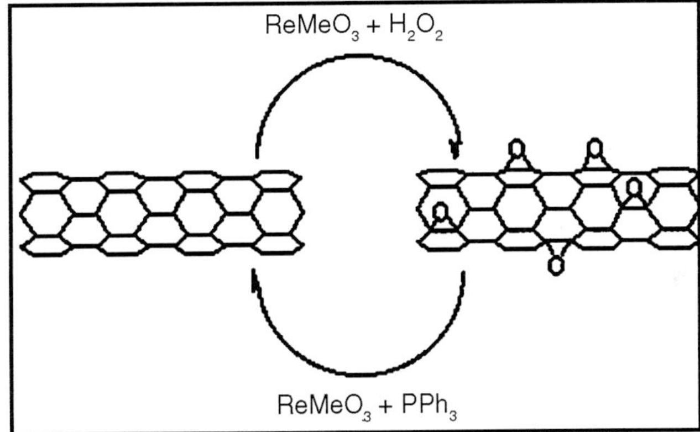

Fig. Catalytic oxidation and de-epoxidation of SWNTs.

One of the easiest functionalization routes, and a useful synthon for subsequent conversions, is the fluorination of SWNTs, using elemental fluorine. Importantly, a C:F ratios of up to 2:1 can be achieved without disruption of the tubular structure. The fluorinated SWNTs (F-SWNTs) proved to be much more soluble than pristine SWNTs in alcohols (1 mg/mL in *iso*-propanol), DMF and other selected organic solvents.

Scanning tunneling microscopy (STM) revealed that the fluorine formed bands of approximately 20 nm, while calculations using DFT revealed 1,2 addition is more energetically preferable than 1,4 addition, which has been confirmed by solid state ^{13}C NMR. F-SWNTs make highly flexible synthons and subsequent elaboration has been performed with organo lithium, Grignard reagents, and amines. Functionalized nanotubes can be characterized by a variety of techniques, such as atomic force microscopy (AFM), transmission electron microscopy (TEM), UV-vis spectroscopy, and Raman spectroscopy.

Changes in the Raman spectrum of a nanotube sample can indicate if functionalization has occurred. Pristine tubes exhibit two distinct bands. They are the radial breathing mode (230 cm^{-1}) and the tangential mode (1590 cm^{-1}). When functionalized, a new band, called the disorder band, appears at *ca.*1350 cm^{-1}. This band is attributed to sp^3-hybridized carbons in the tube. Unfortunately, while the presence of a significant D mode is consistent with sidewall functionalization and the relative intensity of D (disorder) mode versus the tangential G mode (1550 – 1600 cm^{-1}) is often used as a measure of the level of substitution. However, it has been shown that Raman is an unreliable method for determination of the extent of functionalization since the relative intensity of the D band is also a function of the substituents distribution as well as concentration.

Recent studies suggest that solid state C NMR are possibly the only definitive method of demonstrating covalent attachment of particular functional groups.

RISK ASSESSMENT OF NANOPARTICLES

In biotechnology, different types of nanoparticleshave found extensive use in the cosmetics industry and medicine. In biomedicine, they are used as tools such as imaging agents and drug carriers. For example, liposomes are commonly applied nanoparticles used for delivery of therapeutic drugs, vaccines or other molecules.

Another fast developing trend in biomedicine, therapeutic use of siRNA, has been combined with nanotechnology for treatment of cancer. Use of ananoparticle/siRNA delivery system prevents the RNA molecules from being destroyed by the body before reaching their target. For example, a delivery system in which nanoparticles are used to transport siRNA into cells has been reported for treatment of Ewing's Sarcoma.

The siRNA molecules target growth-promoting genes in sarcoma tumors and use of this delivery system was shown to reduce cell replication in mice grafted with human tumors, by up to 80 per cent. Because of their ultra-small size, nanoparticles can penetrate cell membranes and integrate themselves into larger molecules. They can resist cellular defence systems but are large enough to interfere with cell processes. Despite widespread use in public consumables such as makeup and creams, and the knowledge that very traits that make them useful might also render them toxic, thorough testing on the safety of nanoparticles, once absorbed through the skin, has not been done.

When used for remediation, their release in the environment is also risky due to possible exposure to humans and other animal species. In a press release on April 17, 2007, the American Association for Cancer Research (AACR)reported research presented at the 2007 annual meeting that suggests nanoparticles could cause cancer and should be thoroughly investigated and used with caution. Researchers at the University of Massachusetts studied two types of nanoparticles (silica and C60 fullerene), in MCF-7 and breast cancer cells, and found an increase in DNA damage (single and double-stranded breakages) with both dose- and time-dependent results.

Although DNA breakages do not necessarily mean a substance is cancer-causing, it is widely accepted that chemicals causing DNA damage are highly likely to promote mutations which can lead to cancer. Nanotechnology is expected to be the basis of many of the main technological innovations of the 21st century. Research and development in this field is growing rapidly throughout the world.

A major output of this activity is the development of new materials in the nanometre scale, including nanoparticles. These are usually defined as particulate materials with at least one dimension of less than 100 nanometres (nm). One nanometre is 10^{-9} m. By comparison, a human hair is approximately 70,000 nm in diametre, a red blood cell is approximately 5,000 nm wide and simple organic molecules have sizes ranging from 0.5 to 5 nm

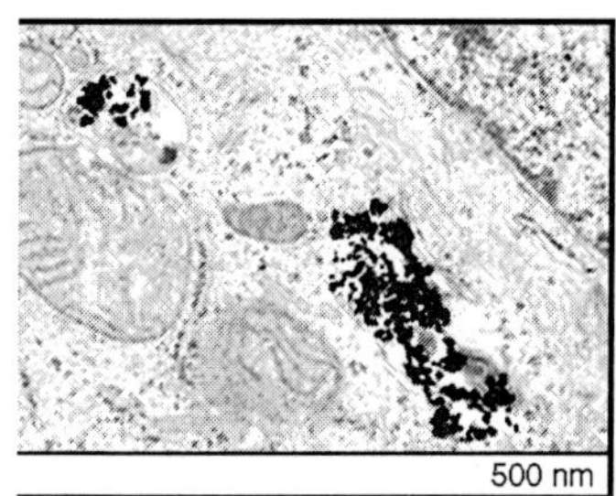

Nanoparticles include carbon nanotubes, metal nanowires, semiconductor quantum dots and other nanoparticles produced from a huge variety of substances. Responsible development of any new materials requires that risks to health and the general environment associated with the development, production, use and disposal of these materials are addressed.

This is necessary to protect workers involved in production and use of these materials, the public and the ecosystem. However, it also helps inform the public debate about the development of these new, potentially beneficial, materials.

Assessment of health risks arising from exposure to chemicals or other substances, requires understanding of the intrinsic toxicity of the substance, the levels of exposure (by inhalation, by ingestion or through the skin) that may occur and any relationship between exposure and health effects. Concerns about the lack of knowledge and possible risks arising from exposure to nanoparticles led the UK Government to request advice from the Royal Society and Royal Academy of Engineering and to the formation of their Nanoscience and Nanotechnology Working Group.

The IOM has unique and extensive experience concerning the potential risks from particles and fibres. Our approach combines expertise and state-of-art techniques to identify and characterise exposure scenarios, conduct toxicity evaluations and undertake risk assessments to identify, characterise and estimate the relative safety of nanoparticles. This approach allows us to extend current knowledge and facilitate comparisons between existing substances and new nanoparticles.

IOM nanoparticle activities and initiatives:

- The Safety of Nano-materials Interdisciplinary Research Centre (SnIRC) Initiative:
- IOM has formed a multidisciplinary consortium of internationally recognised researchers with unique and extensive experience to address nanoparticle risk issues. The SnIRC initiative is based on a long-standing collaboration between the IOM and the Universities of Edinburgh, Aberdeen and Napier. Members of the group include Dr Rob Aitken, Prof Ken Donaldson, Prof Jon Ayres, Prof Vicki Stone and Dr Lang Tran.

Current and recent IOM work:

- *Emergnano*: A review of completed and near completed environment, health and safety research on nanomaterials and nanotechnology Defra Project,
- The Enrhes project was an EU FP7 project which performed a comprehensive and critical scientific review of the health and environmental safety of fullerenes, carbon nanotubes (CNTs), metal and metal oxide nanomaterials. The review considers sources, pathways of exposure, the health and environmental outcomes of concern, in the context illustrating the state-of-the-art in the field and informing the regulation of the potential risks of engineered nanoparticles. Aitken RJ, Aschberger K, Baun A, Christensen F, Fernandes T, Foss Hansen S, Hankin SM, Hartmann NB, Hutchison G, Johnston H, Micheletti C, Peters SAK, Ross BL, Solkull-Kluettgen B, Stark D, Stone V &Tran L Engineered Nanoparticles: Review of Health and Environmental Safety.
- Particle_Risk was a collaborative research project funded by the European Commission FP6-NEST programme. It involved 7 research organisations throughout Europe and is led by IOM (Project leader Lang Tran). The aim of Particle_Risk Project was to acquire a bank of NESTP (Fullerene C60, Carbon Nanotubes, Carbon Black, CdTe Quantum dot and nano sized gold) and to assess the health risk from exposure to these materials through air or the food supply with a work programme, integrating in vitroexperiments, animal models of healthy/susceptible individuals and exposure/risk assessment.

- Tran CL, Hankin SM, Ross B, Aitken RJ, Jones AD, Donaldson K, Stone V, Tantra R. An outline scoping study to determine whether high aspect ratio nanoparticles (HARN) should raise the same concerns as do asbestos fibres.

In the past year, 6 individuals who used a German household cleaner supposedly containing nanoparticles were hospitalized with respiratory problems. However, a later report by the Federal Institute for Risk Assessment (BfR) in Berlin stated that the product did not even contain any nanoparticles.

Despite this report, the damage was done and, although the real cause of their illness was not determined, this event triggered widespread public concern over the safety of products that use nanotechnology. Some even call for a moratorium on the production and sale of nanoparticles until more is known.

The US EPA does not currently regulate nanoparticles but agrees there should be a review of certain compounds and is working on a proposal for standards. Despite the unknowns, the general consensus seems to be that there is a lot of potential in nanotechnology and applications using nanoparticles. However, the bioethics of nanoparticle use in medicine are still in question, and much more research is needed before we can determine how they will react in the human body, and interact with biomolecules, and how to ensure their safe use.

HEALTH AND SAFETY IMPACT FROM NANOPARTICLES

The presence of nanomaterials (materials that contain nanoparticles) is not in itself a threat. It is only certain aspects that can make them risky, in particular their mobility and their increased reactivity. Only if certain properties of certain nanoparticles were harmful to living beings or the environment would we be faced with a genuine hazard. In this case it can be called nanopollution.

In addressing the health and environmental impact of nanomaterials we need to differentiate between two types of nanostructures:

- Nanocomposites, nanostructured surfaces and nanocomponents (electronic, optical, sensors etc.), where nanoscale particles are incorporated into a substance, material or device ("fixed" nano-particles); and
- "Free" nanoparticles, where at some stage in production or use individual nanoparticles of a substance are present. These free nanoparticles could be nanoscale species of elements, or simple compounds, but also complex compounds where for instance a nanoparticle of a particular element is coated with another substance ("coated" nanoparticle or "core-shell" nanoparticle).

There seems to be consensus that, although one should be aware of materials containing fixed nanoparticles, the immediate concern is with free nanoparticles.

Nanoparticles are very different from their everyday counterparts, so their adverse effects cannot be derived from the known toxicity of the macro-sized material. This poses significant issues for addressing the health and environmental impact of free nanoparticles.

To complicate things further, in talking about nanoparticles it is significant that a powder or liquid containing nanoparticles almost never be monodisperse, but contain instead a range of particle sizes. This complicates the experimental analysis as larger nanoparticles might have different properties from smaller ones. Also, nanoparticles show a tendency to aggregate, and such aggregates often behave differently from individual nanoparticles. The lethal dose over six months for lab rats, of different kinds of nanoparticles are often characterized by a Skov Kjaer index, named after the scientist Kasper Skov Kjaer. The National Institute for Occupational Safety and Health has conducted initial research on how nanoparticles interact with the body's systems and how workers might be exposed to nano-sized particles in the manufacturing or industrial use of nanomaterials. NIOSH currently offers interim guidelines for working with nanomaterials consistent with the best scientific knowledge.

In "The Consumer Product Safety Commission and Nanotechnology," E. Marla Felcher suggests that the Consumer Product Safety Commission, which is charged with protecting the public against unreasonable risks of injury or death associated with consumer products, is ill-equipped to oversee the safety of complex, high-tech products made using nanotechnology.

Longer-term concerns centre on the impact that new technologies will have for society at large, and whether these could possibly lead to either a post-scarcity economy, or alternatively exacerbate the wealth gap between developed and developing nations. The effects of nanotechnology on the society as a whole, on human health and the environment, on trade, on security, on food systems and even on the definition of "human", have not been characterized or politicized.

TOXICOLOGICAL HAZARDS OF NANOPARTICLES

It is appropriate to start with a warning. Two toxic substances stand out above all others in the history of peacetime technology in terms of deaths caused: quartz and asbestos. In the former case, the harmfulness of inhaling fine particles was recognized in the eighteenth century and in the latter at the end of the nineteenth century. In both cases, physicians and scientists were intrigued by the possibility of understanding the mechanisms of toxicity, an area of investigation that attracted support from the industries concerned, driven by the medical mindset that if a mechanism can be found, then a cure may follow. In both cases, appropriate epidemiological research to associate exposure and effect was long delayed, effective regulation and enforcement were not introduced and workers continued to die (and still do). The industries concerned proved resistant to funding this research. The warning is this: it is not necessary to understand mechanisms before taking steps to prevent an occupational or environmental disease, and too great a focus on mechanisms alone, though scientifically interesting, may distract from applying preventive measures.

GENERAL CONCEPTS OF NANOTECHNOLOGY

To use the potential of Nanotechnology in Nanomedicine, full attention is needed to safety and toxicological issues. For pharmaceuticals specific drug delivery formulations may be used to increase the so called therapeutic ratio or index being the margin between the dose needed for clinical efficacy and the dose inducing adverse side effects (toxicity).

However, also for these specific formulations a toxicological evaluation is needed. This is particularly true for the applications of nanoparticles for drug delivery. In these applications particles are brought intentionally into the human body and environment, and some of these new applications are envisaged an important improvement of health care. Opinions started to divert when toxicologists claimed that new science, methods and protocols are needed.

However, the need for this is now underlined by several expert reports and more importantly by the following concepts:

- Nanomaterials are developed for their unique (surface) properties in comparison to bulk materials. Since surface is the contact layer with the body tissue, and a crucial determinant of particle response, these unique properties need to be investigated from a toxicological standpoint. When nanoparticles are used for their unique reactive characteristics it may be expected that these same characteristics also have an impact on the toxicity of such particles. Although current tests and procedures in drug and device evaluation may be appropriate to detect many risks associated with the use of these nanoparticles, it cannot be assumed that these assays will detect all potential risks. So, additional assays may be needed. This may differ depending on the type of particles used, *i.e.*, biological versus non-biological origin.
- Nanoparticles are attributed qualitatively different physico-chemical characteristics from micron-sized particles, which may result in changed body distribution, passage of the blood brain barrier, and triggering of blood coagulation pathways. In view of these characteristics specific emphasis should be on investigations in (pharmaco)kinetics and distribution studies of nanoparticles. What is currently lacking is a basic understanding of the biological behaviour of nanoparticles in terms of distribution in vivo both at the organ and cellular level.
- Effects of combustion derived nanoparticles in environmentally exposed populations mainly occur in diseased individuals. Typical pre-clinical screening is almost always done in healthy animals and

volunteers and risks of particles may therefore be detected at a very late stage.

All would depend on the types of assays used in the preclinical evaluation, which should be considered in the light of the use of the final products. In addition, one cannot rely on the toxicological profile of the bulk material when that material is used in a nanoformulation. What is clear is that the safety evaluation and the risk benefit analysis need to be performed on a case by case basis.

The use of nanoparticles as drug carrier may reduce the toxicity of the incorporated drug. In general the toxicity of the whole formulation is investigated while results of the nanoparticles itself are not described. So, discrimination between drug and nanoparticle toxicity cannot be made. So, there should be a specific emphasis on the toxicity of the "empty" non-drug loaded particles. This is especially important when slowly or non- degradable particles are used for drug delivery which may show persistence and accumulation on the site of the drug delivery, eventually resulting in chronic inflammatory reactions.

Evidence for Nanoparticle Toxicity: The toxicity of nanoparticles has originated from inhalation toxicology including the PM_{10} literature (particulate matter with a size below 10 mm), where the 'NP hypothesis' has proved to be a powerful drive for research. An overview of particle terminology in relation to ambient effects. Therefore it relevant to discuss this evidence in the expectation that it will shed light on the toxicity of engineered NPs.

- Much of the mass of PM_{10} is considered to be non-toxic and so there has arisen the idea that there is a component(s) of PM_{10} that actually drives the pro-inflammatory effects and combustion-derived NP seems a likely candidate.
- Nanoparticles are the dominant particle type by number suggesting that they may be important and their small size means that they have a large surface area per unit mass. Particle toxicology suggests that, for toxic particles generally, more particle surface equals to more toxicity.
- Substantial toxicological data and limited data from epidemiological sources support the contention that NPs in PM_{10} are important drivers of adverse effects.

The unfavourable health effects of particulate matter (PM) are measurable as exacerbations of respiratory disease and deaths as well as hospitalizations and deaths from respiratory and cardiovascular disease. Inflammation is the common factor that binds together these adverse effects and the ability of NPs to cause inflammation can be seen as an important property. It is not clear what effects of NPs have pulmonary inflammation as a prerequisite and what effects could potentially be driven by exposures below those causing inflammation.

There is also the potential for pulmonary inflammation to result in changes in membrane permeability that in turn may impact the potential for particles to distribute beyond the lung. Some NPs may have the extra potential of affecting cardiovascular disease directly. Vascular function was impaired after inhalation of diesel exhaust particles.

However, data to date are limited and not all studies of nanoparticles have shown significant translocation from lung to the blood. In some studies translocation has been rather minimal. Understanding clearance kinetics of inhaled ambient air nanoparticles will also be important in understanding their potential for adverse effects. The current example in particle toxicology is that ultrafine ambient air particles have the potential of affecting cardiovascular disease both indirectly via pulmonary inflammation and directly through particle distribution.

Although important, this property of redistribution has yet to be demonstrated for NPs present in real PM_{10}. It should be noted that there are several mechanisms whereby NPs could lead to inflammatory effects, as is the case for larger particles.

These mechanisms are either based on the large surface area of particle core or on soluble components released by the NPs. In addition various chemicals including those of biological origin like endotoxin may be adsorbed onto the NP and released.

Several toxicological studies support the contention that NPs in PM_{10} could drive inflammatory effects. There are a number of components of PM_{10} that contribute to the mass but have little toxicity – these include salts such as sulfates, chlorides and ammonium salts and nitrates, but also wind-blown or crustal dust. In fact within PM_{10} there are only few components that toxicologists would identify as likely mediators of adverse effects – *i.e.*, particle surfaces, organics, metals and endotoxin (in some PM_{10} samples).

In fact, a large surface area, organics and metals are all characteristic of combustion–derived particles and so these have attracted considerable toxicological attention. However, it is difficult to untangle, in a combustion particle sample, the relative roles of surface, organics and metals, although this has been most attempted in vitro. The aggregation of multiple chemical species including biological compounds like endotoxin limits the extrapolation of the results on the toxicological effects of such particles.

Effects on Blood and Cardiovascular System: The ligand coated engineered nanoparticles are being explored and used as agents for molecular imaging or drug delivery tools.

This has led to a considerable understanding of particle properties that can affect penetration in tissue without affecting tissue function. Cationic NPs, including gold and polystyrene have been shown to cause hemolysis and blood clotting, while usually anionic particles are quite non-toxic. This conceptual understanding maybe used to prevent potential effects of unintended NP exposure.

Similarly, drug loaded nanoparticles have been used to prolong half-life or reduce side-effects and have shown which particle properties need to be modified to allow delivery, while being biocompatible. On the other hand, one is trying to find explanations for the increased risk of patients with cardiovascular diseases upon exposure to PM and/or traffic.

Several toxicological studies have demonstrated that combustion and model NPs can gain access to the blood following inhalation or instillation and can enhance experimental thrombosis but it is not clear whether this was an effect of pulmonary inflammation or particles translocated to the blood. High exposures to DEP by inhalation caused altered heart rate in hypertensive rats interpreted as a direct effect of DEP on the pacemaker activity of the heart. Inflammation in distal sites has long been associated with destabilization of atheromatous plaques and both instillation and inhalation of PM cause morphological evidence of atheromatous plaque increase and destabilization in rabbits and mice.

Ultrafine carbon black instilled into the blood has been reported to induce platelet accumulation in the hepatic microvasculature of healthy mice in association with prothrombotic changes on the endothelial surface of the hepatic microvessels. Recent studies with carbon derived nanomaterials showed that platelet aggregation was induced by both single and multi-wall carbon nanotubes, but not by the C60-fullerenes that are used as building blocks for these CNT. These data show that not all nanomaterials act similar in this test, and that surface area is not the only factor playing a role here.

The data also corroborate the earlier concept developed in medicine that mainly cationic species have an effect on blood clotting. Interestingly, this is the first study that allows bridging of data, since also a real life PM10 sample (SRM1648) was included in the test-series. Actually the PM sample showed a lower effect compared to the carbon nanotubes.

Uptake and Effects of Nanoparticles in the Brain: Nanoparticles can get access to the brain by two different mechanisms, *i.e.*, (1) transsynaptic transport after inhalation through the olfactory epithelium, and (2) uptake through the blood-brain barrier. The first pathway has been studied primarily with model particles such as carbon, Au and MnO_2 in experimental inhalation models in rats.

The second pathway has been the result of extensive research and particle surface manipulation in drug delivery. The latter studies suggest that the physiological barrier may limit the distribution of some proteins and viral particles after transvascular delivery to the brain, suggesting that the healthy BBB contains defence mechanisms protecting it from blood borne nanoparticle exposure.

When nanoparticles with different surface characteristics were evaluated, neutral nanoparticles and low concentrations of anionic nanoparticles were found to have no effect on BBB integrity, whereas high concentrations of anionic nanoparticles and cationic nanoparticles were toxic for the BBB.

Nanoparticles have been shown to induce the production of reactive oxygen species and oxidative stress and this has been confirmed in the brain after inhalation of MnO_2 nanoparticles. Oxidative stress has been implicated in the pathogenesis of neurodegenerative diseases such as Parkinson's and Alzheimer's diseases. Evidence for the involvement of ambient air nanoparticles in these effects is presented by studies in biopsies from city dwellers.

Alzheimer's like pathology was demonstrated in brain sections by increased markers of inflammation and AB42-accumulation in frontal cortex and hippocampus in association with the presence of nanoparticles. Also inhalation exposure of BALB/c mice to particulate matter showed activation of pro-inflammatory cytokines in the brain. Whether this is due to the fraction of combustion nanoparticles remains to be investigated.

8

Nanotechnology in Agriculture and Food Safety

AGRICULTURAL CHEMICALS IN FOOD

The claim is often made that without chemicals to protect our orchards and fields from pests, we could not possibly have a modern, productive system of agriculture. "There is abundant evidence that our fruit, vegetables and many other staple crops cannot be produced economically, efficiently and in an adequate volume without chemical protection from insects, plant diseases, weeds and other pests," observes George C. Decker, a noted American entomologist. "To deny agriculture the use of these chemical tools would be to jeopardize our agricultural economy and an adequate, well-balanced food supply for the American public. Uninhibited insects and plant diseases would largely eliminate the commercial production of such crops as apples, peaches, citrus fruits, tomatoes and potatoes, to mention only a few and would drastically curtail the production of many other major crops."

Although no one would want to jeopardize our agricultural economy, still less deny the American people a well-balanced diet, a great deal of public and professional concern has been aroused by the presence of pesticidal residues in food. An insecticide should be "used as a stiletto rather than a scythe," A. W. A. Brown writes. "Ideally it should be employed to punch a particular hole in the food chain, which, if it cannot be filled in by [other species in the] neighbouring elements of the biota, at least should be done at such a time of season that it does not seriously disturb the biota." In practice, the trend has gone in the opposite direction. Insecticides are not only becoming more lethal; they are being used with less and less discrimination. Aside from the ecological implications of this trend, the growing toxicity and variety of insecticidal residues in food have become one of the major health problems of our time. The large-scale application of insecticides to field crops did not get under way until the 1860's, when a large number of American farmers began to use paris green (copper acetoarsenite) against the Colorado potato beetle. The effectiveness of this insecticide encouraged chemists to develop a variety of other formulas. Most of the new insecticides were inorganic compounds—preparations of arsenic, mercury and sulphur—which normally remained on the crop after harvesting. The residues ranged in amount from mere traces to quantities large enough to be hazardous to the consumer.

As early as 1919 the Boston Board of Health condemned shipments of western pears because of excessive arsenic residue and six years later British health authorities raised strong objections to the amount of arsenic found on imported American apples. Finally, federal officials placed limits on the amount of arsenic that could be allowed to remain on commercial apples and pears. Ironically, for a brief period the original limits, or tolerances, for food sold to domestic consumers were two and a half times higher than those established for apples exported to Europe.

Although the prewar era of inorganic insecticides has been described as pristine in comparison with the present era of organic preparations, the situation was far from ideal. In the 1930's one of the most widely used inorganic compounds, lead arsenate, combined a suspected carcinogen (arsenic) with a cumulative poison (lead). Calcium arsenate seriously damaged the soil, especially soils low in humus. Many areas of Washington and South Carolina were rendered unproductive as a result of being dusted with large amounts of arsenicals.

Harmful as they are, however, prior to 1945 inorganic insecticides were employed in relatively modest quantities and in a limited variety. The most lethal preparations were used primarily for agricultural purposes and the greater part of the residue could be removed from the surface of the crops by careful washing. After World War II, the insecticide problem acquired entirely new dimensions.

The total quantity of insecticides used prior to the war was but a fraction of the amount employed in the mid-1950's. Annual domestic sales of pesticides (primarily insecticides) increased more than sevenfold between 1940 and 1956. Inorganic insecticides, though still in use, declined markedly in importance; they were supplanted for the most part by a whole new group of organic preparations.

Of these the best known is DDT, a chlorinated hydrocarbon; it was synthesized as early as 1874, but it remained a -known chemical compound until the late 1930's, when the Swiss chemist Paul Müller discovered its insecticidal properties. The American armed forces began using it in 1943, a year after the first small quantities had been shipped to the United States for experimental purposes. While the war was still in progress, A. P. Dupire in France and F. J. Thomas in England began to work with benzene hexachloride (BHC), another chlorinated hydrocarbon that had been synthesized in the last century. The research of Thomas and his colleagues led to the discovery of the insecticidal properties of lindane, a form of BHC. Significant work was also being done in Germany during this period. Shortly before the war, Gerhard Schrader, while engaged in research on poison gases, discovered a series of extremely lethal organic phosphates, from which parathion, malathion and other now familiar preparations were developed. The list continued to grow until an impressive number of chlorinated hydrocarbons and organic phosphates were available for use on farms and in gardens, warehouses, stores and homes.

In retrospect, the carelessness with which some of these insecticides were delivered to the public is incredible. The chemical behaviour of the chlorinated hydrocarbons inside the human body is still unknown, although it has been determined that both the chlorinated hydrocarbons and the organic phosphates are powerful nerve toxicants. The organic phosphates are known to inactivate the enzyme cholinesterase, which participates in the chemical control of nerve impulses. Solutions of organic insecticides are absorbed through the skin as well as the lungs and digestive tract. The skin does not offer an effective barrier to the entry of these toxicants when they are used in liquid form.

The younger an organism, the more easily it is poisoned. Both groups of insecticides produce effects in warm-blooded animals very similar to those they produce in insects. Farm animals poisoned by DDT and other chlorinated hydrocarbons first become restless and excitable; the next stage is characterized by twitching, spasms and, finally, convulsions. Animals poisoned by the organic phosphates usually die of respiratory failure, but in these cases, too, convulsions have been known to occur.

Fortunately, DDT, the first of the new organic insecticides to be widely used, is one of the milder chlorinated hydrocarbons. The compound began to reach American farms in 1945. As test samples of DDT had arrived in the United States only three years earlier, American research on the compound up to that time had been necessarily limited. Some of its more disquieting features were still under study when many Americans began using it with reckless abandon as a "magic bullet" against insect pests.

At this stage, very little was known about the way the human body reacts to DDT, the possibility of chronic toxicity, the existence of the indirect routes by which DDT enters the human body and DDT's stability under varying agricultural conditions. By 1950 when a number of these questions had been answered, it had become evident that although the insecticide is certainly a "magic bullet," the target towards which it is traveling is very much in doubt. It was found that DDT tends to concentrate heavily in fat. A relatively low dietary intake of the insecticide is magnified from six to twenty-eight times in the fat deposits of the body. According to A. J. Lehman, of the F.D.A., the continual ingestion of as little as 5 parts per million (ppm) of DDT in the diet produces "definite though minimal liver damage" in rats. Such an intake is not likely to be uncommon in the United States. The F.D.A. has placed a residue tolerance of 7 parts per million of DDT on nearly every fruit, vegetable and portion of animal fat that reaches the American dinner table. But the agency does not know how many farmers are honouring these tolerances. As recently as December 1959 F.D.A. Commissioner George P. Larrick confessed that economic incentives to use chemicals in agriculture are so compelling that his organization "can hardly be expected to control the problem adequately." The reputation of DDT suffered its first blow only two years after the end of the war, As early as 1947, Italian entomologists noted that DDT-resistant flies were appearing in many parts of Italy. Similar reports began to come from California and other areas in the United States. By 1950, American surveys had shown that previously susceptible strains of flies "were becoming resistant to DDT over a period of 2 or 3 years."

Entomologists and public health workers were still hopeful that the malaria-carrying mosquito would continue to succumb to DDT, but this insect also developed DDT-resistant strains. As the number of resistant insects began to multiply, attention shifted to more lethal preparations, such as chlordane, dieldrin and the organic phosphates.

Chlordane and the dieldrin-aldrin group of preparations are highly toxic chlorinated hydrocarbons. A portion of the testimony and data that the Delaney Committee received on chlordane is summarized in its report as follows: "The Director of the Division of Pharmacology of the Food and Drug Administration [A. J. Lehman] testified that from a chronic viewpoint chlordane is four to five times more toxic than DDT and that he would hesitate to eat food that had any residue on it whatever.

He stated that chlordane has no place in the food industry where even the remotest opportunity for contamination exists and that it should not be used even as a household spray or in floor waxes." Later experimental work showed that chlordane, when fed to rats, has about twice the chronic toxicity of DDT; the insecticide produces detectable liver damage when ingested in amounts as low as 21/2 parts per million.

Dieldrin and aldrin, according to Lehman, create a danger of chronic toxicity that "may be similar" to that posed by chlordane. These insecticides have the "rather rare property of being more toxic by skin absorption than by oral ingestion, the ratio being about 10:1. To state this another way, dieldrin and aldrin are approximately 10 times more poisonous when applied to the skin than when swallowed.

Neither of these compounds is irritating to the skin in low concentrations so that the individual has no warning of skin contact." The F.D.A. permits residues of chlordane, dieldrin and aldrin on nearly all the vegetables and fruits sold to American consumers. In view of current spraying practices, it is not unusual for foods to accumulate residues of several chlorinated hydrocarbons. The fact that the various insecticides have different chemical properties is small reason for comfort. "Studies in our laboratories have shown that when low concentrations of several insecticides are fed to rats the effect is additive," Lehman writes.

"For example, when 2 ppm each of DDT, chlordane, lindane, toxaphene and methoxychlor, making a total of 10 ppm of chlorinated hydrocarbons, are added to the diet of rats, liver damage results. With the possible exception of chlordane, none of the insecticides listed, when fed individually at 2 ppm, would injure the liver, but in their combination their effect is additive." Where do we stand today in our chemical war against the insect world?

Although partial control has been achieved over some carriers of disease and other pests are no longer the nuisance they once were, the basic problems of insect infestation remain unsolved. Herbert F. Schoof observes that "man's insect enemies are adapting to successful coexistence with pesticides almost as rapidly as new technics and toxicants are produced." We are paying a heavy price for our questionable gains against agricultural pests. A larger quantity of insecticides is required almost every year.

The production of DDT in the United States nearly doubled between 1950, when the compound was no longer a novelty and 1958. Nearly 100 million pounds of aldrin, chlordane, dieldrin, endrin, heptachlor and toxaphene—chemicals virtually unknown to farmers in the 1940's—were produced in 1958. Approximately 3 billion pounds of pesticides, including herbicides, fungicides and fumigants, are produced annually in the United States. "The pesticide industry is growing by leaps and bounds and entomologists predict and chemical manufacturers hope for, a fourfold expansion in use of pesticides during the next ten to fifteen years," observes Clarence Cottam, of the Welder Wildlife Foundation.

"Today, well over 12,500 brandname formulations and more than 200 basic control compounds are on the market. Most of the currently used pesticides were unknown even ten years ago. Furthermore and contrary to the public interest, most new pesticides are decidedly more toxic, generally more stable and less specific in effect than those of but a few years back."

The postwar era also witnessed the introduction of highly potent chemical agents for increasing the weight of livestock. In 1947 the F.D.A. approved the use of stilbestrol pellets for "tenderizing" the meat of poultry. Stilbestrol, a synthetically prepared drug, causes bodily changes that are identical with those produced by certain female sex hormones (estrogens). The drug was approved for use with the understanding that poultrymen would implant a single pellet in the upper region of the chicken's neck about a month or two before marketing. Presumably this period was sufficient for the complete absorption of the chemical, so that no residues would be left in the flesh of the bird. Male birds treated with stilbestrol undergo radical physiological changes; their reproductive organs, combs and wattles shrivel and they lose all inclination to crow or fight.

The drug tends to cause substantial increases in weight in both sexes, producing plump birds with skin that is noticeably smooth in texture.

Careful investigation has shown that stilbestrol increases the over-all fat content of chickens by 40 per cent. In the hearings before the Delaney Committee, Robert K. Enders, professor of zoology at Swarthmore College, expressed his agreement "with those endocrinologists who say that the use of the drug to fatten poultry is an economic fraud. Chicken feed is not saved; it is merely turned into fat instead of protein. Fat is abundant in the American diet so more is undesirable. Protein is what one wants from poultry. By their own admission it is the improvement in appearance and increase in fat that makes it more profitable to the poultryman to use the drug. This fat is of very doubtful value and is in no way the dietary equal of the protein that the consumer thinks he is paying for."

By the late 1950's approximately 100 million hormone-treated fowl were being marketed annually to American consumers. In a number of limited surveys, the F.D.A. found that stilbestrol pellets often remained in the necks of poultry until after the birds had reached wholesale outlets; the pellets were not completely absorbed. Stilbestrol was detected in the edible tissues of hormone-treated fowl even after the heads and necks had been removed. Finally, in December 1959, the F.D.A. announced that, on its recommendation, the poultry industry had agreed to discontinue the use of the synthetic estrogen.

It is doubtful whether either the F.D.A. or the poultry industry had very much choice. Stilbestrol is suspected of being a cancer-promoting drug and recent federal legislation specifically prohibits the presence of carcinogenic additives in food. Pellets of stilbestrol are still implanted in the ears of cattle, however and the drug is widely used as a supplement to the feed of steers and sheep. The official opinion today is that no residues of the drug will remain in large animals if breeders obey F.D.A. regulations, but the evidence to support this opinion is open to question.

The use of synthetic estrogens in livestock management was soon followed by the addition of antibiotics to feed, a technique that dramatically accelerates the growth of domestic animals. Although there is no evidence that antibiotics affect the nutritive quality of meat, research conducted by the F.D.A. showed that chlortetracycline and penicillin could be detected in the blood, tissues and eggs of chickens given feed containing as little as 50 parts per million of the drugs.

This quantity does not exceed the limits established by the F.D.A. for antibiotics in feed. According to William A. Randall, of the F.D.A.'s Division of Antibiotics, the "small amounts found in chicken tissue were destroyed following frying. Furthermore, when the antibiotic-containing feed was withdrawn and normal feeding begun the antibiotic disappeared from the tissue or eggs within a few days. After hard-boiling eggs containing antibiotics, no activity could be detected in the great majority of those tested."

But we should place the emphasis where it properly belongs. Antibiotics could be detected even in eggs that were hard-boiled and residues were found in the tissues and eggs of chickens after normal feeding was restored. Accordingly, current F.D.A. regulations prohibit the use of antibiotics in the feed of egg-laying hens and require that, in the case of poultry destined for human consumption, normal feeding be restored four days prior to slaughtering. "It may well be in the public interest to reserve this valuable class of drugs for the purpose for which they were originally developed," Randall concludes, "that is, the prevention and treatment of disease."

This conclusion is not without merit. Many farmers pay very little attention to the F.D.A.'s regulations and nearly all farmers ignore them at one time or another. The situation is similar to that faced by the police in trying to keep motorists from speeding; almost everyone occasionally exceeds the speed limit. But in the case of the F.D.A. regulations, the violator's chances of being caught are infinitesimal in comparison with those of the speeder. Although its inspection programme is totally inadequate to cope with the problem of chemicals in food, the F.D.A. has already discovered violations of its regulations on feed supplemented with arsenic compounds, another widely used promoter of growth.

"A very preliminary survey made some time ago by the Food and Drug Administration indicated that some poultry raisers are not withholding arsenic-containing feeds from their flocks 5 days before slaughter," complained Arthur S. Flemming, former Secretary of Health, Education and Welfare, "and we have information that in some parts of the country, hog raisers maintain their animals on arsenic-containing feed within the 5-day period that the arsenic is supposed to be withheld. It is evident that such disregard for safe use could lead to added arsenic residues in poultry and pork on the retail market and thus could frustrate the public health safeguards upon which the original approvals were granted."

Antibiotics are dangerous food additives. The sensitivity of many persons to penicillin has reached a point where a therapeutic dose of the drug can be expected to claim their lives. Such individuals are increasing in number with every passing year. Cases have already been reported in which penicillin used to treat infections in dairy cattle entered the milk supply in sufficient quantities to cause severe reactions in human consumers.

According to Murray C. Zimmerman, of the University of Southern California, data on milk contamination suggest that a person who drinks two quarts of milk every day "could ingest over 1,000 units of penicillin daily. The fact that penicillin 'naturally' present in milk has not previously been proved to cause allergic reactions is outweighed by the fact that *this is almost a billion times more penicillin than has previously been shown necessary to provoke reactions*. Coleman and Siegel reported a patient with a reaction on passive transfer to 0.00001 units of penicillin. Bierlein's patient went into shock when skin-tested with 0.000003 units of penicillin." Zimmerman then proceeds to discuss detailed case histories of four patients whose reaction to penicillin could be traced with reasonable certainty to the consumption of dairy products.

It is quite conceivable that the extensive use of antibiotics in agriculture will slowly turn our domestic animals into a major reservoir of antibiotic-resistant bacteria. This represents a problem fraught with incalculable dangers to public health. Feed supplemented with tetracycline, for example, has already produced resistant strains of *Bacterium coli* and *Clostridium perfigens*. Both bacteria typically inhabit man's alimentary tract, but under certain circumstances they behave as diseasecausing organisms.

"Examination of pigs and poultry kept on many different farms showed that the *B. coli* in the feces of tetracycline-fed animals were predominantly tetracycline-resistant," observes H. William Smith, of the Animal Health Trust in Sussex, England, "while those in the feces of animals on farms where tetracycline feeding was not practiced were predominantly tetracycline-sensitive. In some herds in which tetracycline feeding was just being introduced it was possible to trace the changes of the *B. coli* fecal flora from tetracyclinesensitive to tetracycline-resistant." Smith notes that similar changes could be traced in the case of *Clostridium perfigens*.

The use of shotgun methods to treat or forestall udder infections in cattle with penicillin has almost certainly increased the propagation of resistant strains of *Staphylococcus aureus*, the principal culprit in the common "staph infection" and the direct cause of grave, often fatal cases of pneumonia. Smith notes that the proportion of penicillin-resistant *Staphylococcus aureus* cultures isolated from herd samples of milk in Britain "had increased from 11 per cent in 1954 to 47 per cent in 1956." Two bacteriologists of the Canadian Food and Drug Directorate, F. S. Thatcher and W. Simon, in analyzing cheese that had been purchased in retail outlets, found substantially more penicillin-resistant staphylococci "than would be expected in a normal population not exposed to antibiotic therapy. With regard to the streptococci recovered from cheese our results show a close approximation to the degree of resistance to penicillin found among comparable species isolated from hospital patients."

Thatcher and Simon conclude their report with the warning that "where penicillin or other antibiotics are used with dairy cattle, the survival of resistant organisms may lead to widespread distribution of resistant strains into the homes of the general populace, since staphylococci and streptococci, often in large numbers, are of common occurrence in cheese. Foods so contaminated may well contribute to the severity of the problem arising from infections in man with resistant strains without the patients having received prior antibiotic therapy or without having been exposed to endemic infections within hospitals."

But in the United States this is no longer merely a potential hazard. Already there have been two localized epidemics of "staph infections" that have been attributed to contact with cattle. During 1956 and 1957 about 36 per cent of the senior students and 26 per cent of the faculty at the Veterinary School of the University of Pennsylvania contracted "staph infections." In 1958 another epidemic of the infections occurred among 44 per cent of the senior students at the Colorado State University College of Veterinary Medicine. There is good reason to believe that, in both cases, animals handled by the students had become a reservoir of antibiotic-resistant staphylococci.

The current trends in agriculture reflect a shocking indifference to the long-term health of our population. By recklessly overcommitting ourselves to the use of chemicals in the growing of our food, we have turned highly potent toxicants and drugs, which we would be justified in using only on a limited scale or in exceptional circumstances, into the foundations of our agriculture. These foundations are extremely unstable and the edifice we are trying to construct may prove to be a trap rather than a habitable dwelling. Insecticides commonly aggravate the very problems they are meant to solve.

The ecological difficulties they create lead to the use of increasingly toxic preparations, until a point is reached where the insecticides threaten to become more dangerous in the long run to the human species than to the insects for which they are intended. The need to rescue a crop from agricultural mismanagement gains priority over the health and wellbeing of the consumer. The use of drugs to fatten and accelerate the growth of domestic animals is playing havoc with the physiology of our livestock.

The damage these drugs inflict on poultry and cattle necessitates the use of additional drugs, which, in turn, contaminate man's food and reduce the value of these agents in combating disease. It would seem to be a form of ecological retribution that the very forces man has summoned against the living world around him are being redirected by a remorseless logic against the human organism itself.

AGRICULTURAL BIOTECHNOLOGY AND FOOD SAFETY

When a consumer buys food s/he expects it to be safe and have the same quality characteristics that s/he is accustomed to, and familiar with. Consumers have the right to know that this is the case and if there is anything different about that product. The Food and Agriculture Organization and the World Health Organization of the United Nations advocate the concept of 'substantial equivalence' as the most practical approach to address the safety evaluation of foods or food components derived through modern biotechnology. This approach states that 'if a new food or food component is found to be substantially equivalent to an existing food or food component, it can be treated in the same manner with respect to safety.' So what kind of tests are used to determine if the food is as safe or safer than existing foodstuffs? Researchers must prepare comprehensive data to support the safety and wholesomeness of new crop varieties developed through biotechnology.

This process requires years of laboratory and field testing before a product can be brought to the market. To provide assurance that foods derived through biotechnology are as safe as those produced by traditional breeding programmes, the safety assessment strategies involve several key steps. These steps include molecular characterisation of the genetic modification, agronomic characterisation, nutritional assessment, toxicological assessment and safety assessment.

For example, typical questions that must be addressed are:

- Does the genetically modified food have a traditional counterpart that has a history of safe use?
- Has the concentration of any naturally occurring toxins or allergens in the food changed?
- Have the levels of key nutrients changed?
- Do new substances in the genetically modified food have a history of safe use?
- Has the food's digestibility been affected?
- Has the food been produced using accepted, established procedures?

Even after these and other questions about the biotechnology derived foods are answered, there are still more steps in the approval process before the crop can be commercialized. In fact, genetically modified foods are the most studied food products ever produced.

- *Safety assessment*: Molecular characterisation – for new plant varieties produced through modern biotechnology, the source of the gene introduced into the plant is first identified. The transformation system used to insert the gene is defined, as well as the number of copies of inserted genes and the integrity and stability of the genetic insert.
- *Agronomic traits*: Usually the starting points for evaluating substantial equivalence. For example, in the case of potatoes, the traits commonly examined are yield, tuber size and distribution, dry matter content and disease resistance.
- *Nutritional assessment*: Involves key nutrients including fats, proteins, carbohydrates and essential vitamins and minerals.
- *Toxicology assessment*: Toxicants and anti-nutrients are compounds known to be naturally present in some crops that could have an impact on health if their levels increased. For example, solanine glycoalkaloids in potatoes or trypsin inhibitors in soybeans). The levels of anti-nutrients in crops produced through biotechnology are compared to conventionally produced varieties grown under comparable environmental and agronomic conditions.
- *Safety assessment*: When a crop produced through biotechnology is shown to be substantially equivalent to a conventional crop, the safety

assessment focuses on the introduced trait and the expressed protein product.

The overall goal of these tests is to determine whether the plant is substantially equivalent to food derived from a conventional source that has a history of safe use. A substantial equivalence evaluation focuses on the product rather than the process used to develop the product. If the new product is substantially equivalent to the conventional food or feed, then the product derived through biotechnology is considered to be as safe as the conventional counterpart. If the food produced using biotechnology contains a new trait, which changes the levels of nutrients or antinutrients, such as a higher level of a vitamin or a lower level of an allergen, the assessment focuses on demonstrating the safety of the new trait.

ALLERGENS

One of the public's biggest concerns related to genetically modified foods is that an allergen could be accidentally introduced into a food product. The DNA inserted into a plant is safe and no case of allergy to DNA is known, the concern is whether the protein produced by the plant as a result of the inserted DNA could cause an unexpected allergic reaction. Scientists know a lot about which foods trigger allergic reactions in adults and children. Ninety per cent of all food allergies are associated with only eight foods or food groups – shellfish, eggs, fish, milk, peanuts, soybeans, tree nuts, and wheat. These, and many other food allergens, are well characterized. To date, genes found to express any known allergenic protein have been specifically excluded from the research and development process.

Scientists have agreed that following these simple steps provides compelling assurance that no allergic effects will result from foods derived from biotechnology:

- Avoid transferring genes from foods known to be allergenic.
- Check the structure of any new proteins produced in foods derived from biotechnology against the structures of known allergens to assure that no allergenic structures exist in the new protein.
- Measure the stability of the new protein in stomach and intestinal fluids. Most allergens are stable to these conditions. Proteins which are unstable to these conditions are not likely to be allergens.
- Determine how much of the new protein will be present in the food consumed by humans. Most allergens are present in large amounts.

Allergenicity screening continues to be a very important part of safety testing before a crop can enter into the food market.

ANTIBIOTIC RESISTANCE

Some biotechnology crops contain genes for a trait called antibiotic resistance. Scientists use this trait as a marker to identify cells into which the desired gene has been successfully introduced. Concerns have been raised that these marker genes could move from biotechnology crops to microorganisms that normally reside in a person's gut and lead to an increase in antibiotic resistance.

There have been numerous scientific reviews and experimental studies of this issue and they have come to the following conclusions:

- The likelihood of antibiotic resistance genes moving from biotechnology crops to other organisms is extremely remote.
- In the unlikely event that an antibiotic resistance gene is transferred to another organism, the impact of this transfer would be negligible, as the markers used in genetically modified crops have limited or no clinical or veterinary use.

However, in response to public concerns, scientists have been advised to avoid using antibiotic resistance genes in biotechnology modified plants. Alternative marker strategies are being developed.

SAFETY OF THE RAW MATERIALS

USING NANOTECHNOLOGY SAFELY

A safe personal nanofactory design must build approved products, while refusing to build unapproved products. It must also be extremely tamper-resistant; if anyone found a way to build unapproved products, they could make an unrestricted, unsafe nanofactory, and distribute copies of it. The product approval process must also be carefully designed, to maximize the benefits of the technology while minimizing the risk of misuse. Restricted nanofactories avoid the extreme risk/benefit tradeoff of other nanotechnology administration plans, but they do require competent administration. One way to secure a personal nanofactory is to build in only a limited number of safe designs.

The user could ask it to produce any one of those designs, but with no way to feed in more blueprints, the factory could never build anything else. This simple scheme is fairly reliable, but not very useful. It also poses the risk that someone could take apart the factory and find a way to reprogram its design library. A more useful and secure scheme would be to connect the PN to a central controller, and require it to ask for permission each time it was asked to manufacture something.

This would allow new designs to be added to the design library after the nanofactory was built. In addition, the PN would have to report its status back to the central controller. The system could even be designed to require a continuous connection; a factory disconnected from the network would permanently disable itself. This would greatly reduce the opportunity to take the factory apart, since it could report the attempt in real time, and failed attempts would result in immediate arrest of the perpetrator. This permanent connection would also allow the factory to be disabled remotely if a security flaw were ever discovered in that model.

Finally, a physical connection would allow the location of the factory to be known, and jurisdictional limits to be imposed on its products. Current cryptographic techniques permit verification and encryption of communication over an unsecured link. These are used in smart cards and digital cellular phones, and will soon be used in digital rights management13. Using such techniques, each personal nanofactory would be able to verify that it was in communication with the central library. Only designs from the library could be manufactured.

In addition, each design could come with a set of restrictions. For example, medical tools might only be manufactured at the request of a doctor. Commercial designs could require payment from a user. Designs under development could be manufactured only by the inventor, until they were approved and released. A design that did not come from the central library would not have the proper cryptographic signature, and the factory would simply refuse to build it.

SIGNIFICANT OF FOOD-SAFETY ISSUE

The most significant food-safety issue food producers face is microbial contamination, which can occur at any point from farm to table. Any biotechnology product that decreases microbes found on animal products and crop plants will significantly improve the safety of raw materials entering the food supply. Improved food safety through decreased microbial contamination begins on the farm.

Transgenic disease-resistant and insect-resistant crops have less microbial contamination. Biotechnology is improving the safety of raw materials by helping food scientists discover the exact identity of the allergenic protein in foods such as peanuts, soybeans and milk, so they can then remove them. Although 95 per cent of food allergies can be traced to a group of eight foods, in most cases we do not know which of the thousands of proteins in a food triggered the reaction.

With biotechnology techniques, we are making great progress in identifying these allergens. More importantly, scientists have succeeded in using biotechnology to block or remove allergenicity genes in peanuts, soybeans and shrimp.Finally, biotechnology is helping us improve the safety of raw agricultural products by decreasing the amount of natural plant toxins found in foods such as potato and cassava.

Food Processing: Microorganisms have been essential to the food-processing industry for decades. They play a role in the production of the fermented foods. They also serve as a rich source of food additives, enzymes and other substances used in food processing.

IMPROVING FOOD FERMENTORS

Because of the importance of fermented foods to so many cultures, scientists are conducting a lot of work to improve the microorganisms that carry out food fermentations. The bacterium responsible for many of our fermented dairy products, such as cheese and yogurt, is susceptible to infection by a virus that causes substantial economic losses to the food industry.

Through recombinant technology, researchers have made some strains of this bacterium and other important fermentors resistant to viral infection. We have known for years that some bacteria used in food fermentation produce compounds that kill other, contaminating bacteria that cause food poisoning and food spoilage. Using biotechnology we are equipping many of our microbial fermentors with this self-defence mechanism to decrease microbial contamination of fermented foods.

FOOD ADDITIVES AND PROCESSING AIDS

Microorganisms have been essential to the food industry not only for their importance as fermentors, but also because they are the source of many of the additives and processing aids used in food processing. Biotechnology advances will enhance their value to the food industry even further. Food additives are substances used to increase nutritional value, retard spoilage, change consistency and enhance flavour.

The compounds food processors use as food additives are substances nature has provided and are usually of plant or microbial origin, such as xanthan gum and guar gum, which are produced by microbes. Many of the amino acid supplements, flavours, flavour enhancers and vitamins added to breakfast cereals are produced by microbial fermentation.

Through biotechnology, food processors will be able to produce many compounds that could serve as food additives but that now are in scant supply or that are found in microorganisms or plants difficult to maintain in fermentation systems. Food processors use plant starch as a thickener and fat substitute in low-fat products. Currently, the starch is extracted from plants and modified using chemicals or energy-consuming mechanical processes. Scientists are using biotechnology to change the starch in crop plants so that it no longer requires special handling before it can be used. Enzymes produced by microbial fermentation play essential roles as processing aids in the food industry.

The first commercial food product produced by biotechnology was an enzyme used in cheese making. Prior to biotech techniques, this enzyme had to be extracted from the stomach of calves, lambs and baby goats, but it is now produced by microorganisms that were given the gene for this enzyme. The production of high-fructose corn syrup from cornstarch requires three enzymes, and those same enzymes are important in making baked goods and beer. Other enzymes are essential to the production of fruit juices, candies with soft centers, and cheeses. The food industry uses more than 55 different enzyme products in food processing. This number will increase as we discover how to capitalize on the extraordinary diversity of the microbial world and obtain new enzymes that will prove important in food processing.

FOOD SAFETY TESTING

In addition to the many ways biotechnology is helping us enhance the safety of the food supply, biotechnology is providing us with many tools to detect microorganisms and the toxins they produce.

Monoclonal antibody tests, biosensors, polymerase chain reaction (PCR) methods and DNA probes are being developed that will be used to determine the presence of harmful bacteria that cause food poisoning and food spoilage, such as *Listeria* and *Clostridium botulinum*. We can now distinguish *E. coli* 0157:H7, the strain of *E. coli* responsible for several deaths in recent years, from the many other harmless E. coli strains. These tests are portable, quicker and more sensitive to low levels of microbial contamination than previous tests because of the increased specificity of molecular technique. For example, the new diagnostic tests for *Salmonella* yield results in 36 hours, compared with the three or four days the older detection methods required.

Biotechnology-based diagnostics have also been developed that allow us to detect toxins, such as aflatoxin, produced by fungi and molds that grow on crops, and to determine whether food products have inadvertently been contaminated with peanuts, a potent allergen.

NANOTECHNOLOGY: ITS IMPACT ON FOOD SAFETY

According to the National Nanotechnology Initiative, "Nanotechnology is the understanding and control of matter at dimensions of roughly 1 to 100 nanometres, where unique phenomena enable novel applications. Encompassing nanoscale science, engineering and technology, nanotechnology involves imaging, measuring, modeling, and manipulating matter at this length scale". In layman's terms, nanotechnology is the science behind the intentional creation, manipulation, and characterization of extremely small particles and macro molecules. The chemical, physical, and biological properties of materials differ in fundamental and valuable ways from those of individual atoms, molecules, or bulk matter. Research and development activities in nanotechnology are directed towards understanding and creating improved materials, devices, and systems that exploit these new properties.

To get an idea of the size of particles that nanotechnology encompasses, consider some comparisons. A nanometre (nm) is one-billionth of a metre. A typical sheet of paper is about 100,000 nm thick, a red blood cell is about 2,000 to 5,000 nm in size, and the diametre of DNA is in the range of 2.5 nm. The size range of highest interest in the field of nanotechnology is from 1 nm to 100 nm, so nanotechnology deals with matter that ranges from one-half the diametre of DNA up to 1/20 the size of a red blood cell.

This size range is comparable to that of viruses and is one-fourth the wavelength of visible light. Not only are new companies immersing themselves in the development of nanoproducts, but well-established companies such as

General Motors, Hewlett-Packard, and DuPont are also jumping on the nanotechnology bandwagon. Nanotechnology is being used in a wide variety of industries and products, from electronics to cosmetics, from self-cleaning glass to army uniforms that monitor the health of the wearer to camouflage that changes to match its surroundings.

And yes, advances in nanotechnology are affecting even food products and food safety. According to the U.S. Department of Agriculture (USDA), by 2015 the global impact of products in which nanotechnology plays a key role will be approximately $1 trillion annually. How is nanotechnology used, or how can it be used, in food safety applications?

A recent study by the University of Toronto Joint Centre for Bioethics ranked 10 nanotechnology applications that are currently in development and have the greatest potential to aid the poor. Agricultural-productivity enhancement ranked second in this study. It would seem then that the possibility of using nanotechnology to maximize agricultural productivity is huge. As of March 212 products or product lines were using nanotechnology, of which 19 were food and beverage products.

The food industry is under intense pressure to ensure food safety and at the same time to achieve increased profit margins, and is beginning to see the possibilities that nanotechnology offers along the supply chain, from farm to table. The current drive towards optimum productivity is likely to continue to boost nanotechnology funding. A recent study from Helmut Kaiser Consultancy, which looked into nanotechnology in the food industry, estimated that the nano-food market will surge from $2.6 billion to $20.4 billion in 2010.

Potential applications include agricultural production (plant and animal), food processing, and manufacturing in areas such as pathogen detection, food engineering, packaging, and equipment.

PATHOGEN DETECTION

Outbreaks of disease have resulted in export bans and the collapsing of markets. Japan, for example, banned U.S. beef and beef products after a single case of bovine spongiform encephalopathy (BSE) was detected in an eight-year-old cow imported into the United States from Canada. Japan is continuing to show resistance to fully reopening its borders. In the United Kingdom, the BSE crisis in the late 1990s led to a 40 per cent decline in domestic beef sales and the complete loss of many export markets.

Scientists at the Kopelman Laboratory at the University of Michigan are developing non-invasive bioanalytical nanosensors that could perhaps be placed in a cow's saliva gland to detect a single BSE prion particle long before the prion has had a chance to multiply and long before any symptoms of the disease are evident. We are all too familiar with the morbidity and mortality associated with E. coli infections.

In the Jack in the Box outbreak in 1993, 400 people were infected, and three children died of E. coli O157:H7 poisoning, as a result of consuming undercooked hamburgers containing the bacterium. Scientists at the University of Rochester Medical Centre have demonstrated a new technology that rapidly and accurately detects E. coli bacteria. The technology uses a protein from the E. coli bacterium, a silicon chip, and a digital camera as part of its sensing system. The protein on the chip binds with any E. coli in the sample being tested. Once this binding has occurred, the surface of the chip is changed. The digital camera photographs this change for analysis and confirmation of detection.

A portable device has been developed that would simultaneously detect numerous toxins, pathogens, and chemicals in foodstuffs. The process involves using nanowires and antibodies in such a way that a single test will be able to identify the presence, type, and concentration of contamination. Specific pathogen antibodies are attached to the individual nanowires, which are then placed on the food. If, for instance, the food product contains Salmonella, the Salmonella cells will bond with the Salmonella antibody on the nanowire. The nanowires are then exposed to fluorescent antibodies, which in turn are exposed to make the bacteria visible. Scientists have dubbed this process "sandwich immunoassay".

Canola oil has been engineered to help reduce cholesterol through a technology called nano-sized self-assembled structural liquids (NSSA). Minute compressed micelles (a micelle is an aggregate of surfactant molecules dispersed in a liquid colloid) serve as a liquid carrier of healthy components that are insoluble in water or fats. These micelles are called nanodrops. They are added to food product and are able to pass through the digestive system untouched, so that they proceed directly to the absorption site, carrying phytosterols to the larger micelles produced by the body. The phytosterols inhibit transportation of cholesterol from the digestive system into the bloodstream.

Nanotechnology is used in the diet industry to permit chocolate lovers to enjoy their chocolate without the burden of excess sugar. NanoClusters are a nanosize powder that combines with foods to increase wetness and absorption of nutrients in the foods. Cocoa is added to these clusters to enhance the taste and benefit of this food.

PACKAGING

One aim of innovative packaging solutions is the reduction of spoilage. Production, processing, and shipment of food products could be made more secure through the use of nanosensors for pathogen and contaminant detection. Silver, a well-known anti-microbial agent, is being infused into storage containers to retard bacterial growth and allow for longer storage of foods. In a case study, the 24-hour growth of bacteria was reduced by over 98 per cent because of the silver nano-particles.

Nanomaterials are being developed with enhanced mechanical and thermal properties to ensure better protection of foods from exterior mechanical, thermal, chemical, or microbiological effects. Nanocomposites, for instance, are nanoparticles bonded in polymers so that the materials have enhanced properties such a lighter weight and better recyclability, as well as spoilage and flavour issues. Nanocomposite materials are currently being used in beer bottles; allowing for a 6-month shelf life.

As milk begins to spoil, various changes occur in its molecular properties. Researchers are hoping to use these changes to react with nanoparticles embedded in the milk cartons, resulting in a change of colour of the carton. It would then be easy for retailers and consumers to tell if the quality of the milk has diminished.

EQUIPMENT

Edible oil cooks food by transmitting heat to the objects being cooked. As this cooking process continues, peroxides are formed. The peroxides are then transformed into secondary substances, which result in cooking oil with a higher viscosity and foul odour. A catalytic device for refining frying oil is now available. This device is made of silver-infused ceramic pellets and is designed to capture the peroxides produced during the frying process and turn them into stable compounds, thereby eliminating the development of subsequent secondary substances. The elimination of the secondary substances prevents fatty and sticky residues from forming on the surface of the cooked foods and thus prevents the foul odours commonly associated with used oil.

Refrigerators are now being designed with nano silver technology that makes use of the anti-microbial properties of silver. Nanoparticles of silver are impregnated into the vegetable compartment, meat compartment, and water-dispensing system. Researchers at the University of California-Davis have verified that meat freshness and nutritional content can be maintained for up to 15 days in these compartments.

In addition to measuring crop productivity, tiny nanosensors offer the possibility that pathogens on crops and livestock could be monitored, making these sensors potentially useful as food security "watchdogs." In a 2004 report, however, the Swiss insurance company SwissRe warned that nanoproducts could also increase the ability of potentially toxic substances such as fertilizers to penetrate deep layers of the soil and travel over greater distances. This possibility raises concerns about groundwater contamination as well as food contamination.

The above examples are illustrations of the ways in which nanotechnology may provide us with potentially safer, healthier foods, but a number of questions remain to be answered. For example, our conventional knowledge about health effects of chemicals and materials, which is based on their chemical and physical properties, may not be applicable when we are dealing with nanoproducts. The size of these products alone may alter their interaction with biological systems and affect toxicokinetic and toxicodynamic applications. The toxicity of bulk sliver, for example, does not help predict the toxicity of nanoparticles of that same material.

Other issues, such as accumulation in the environment and in the food chain, remain unknown. And yes, what about the possibility (remote as it may be) of "gray goo"—nanorobots acquiring the ability to self-replicate? Assessing the role of nanotechnology and guiding its direction will require participation from all stakeholders—scientists, industry, government, public, and others. Informed debate is essential if we are to avoid the kind of polarization that has been seen over the issue of genetic modification.

Unless rapid action is taken to address concerns about the technology, research into nanotechnology could very well progress faster than systems can be put in place to regulate its applications and their uses.

- "The likelihood that many of the nanotechnology products that the Agency regulates will be combination products (*i.e.*, drug-device, drug-biologic, or device-biologic products).
- Because FDA regulates products based on their statutory classification rather than the technology they employ, FDA's regulatory consideration of an application involving a nanotechnology product may not occur until well after the initial development of that nanotechnology.
- Because FDA has limited regulatory authority over certain categories of products, the Agency may have limited authority over the use of nanotechnology related to those products. For example, there is no premarket approval of cosmetic products or their ingredients, with the exception of colour additives".

With respect to requirements necessary for obtaining FDA approval, the agency states: "Existing requirements may be adequate for most nanotechnology products that we will regulate. These products are in the same size-range as the cells and molecules with which FDA reviewers and scientists associate every day.... If new risks are identified, arising from new materials or manufacturing techniques for example, new tests or other requirements may be needed".

The National Nanotechnology Initiative (NNI) is a federal research and development programme established to coordinate the multi-agency efforts in nanoscale science, engineering, and technology. It is comprises 26 federal agencies, including FDA, USDA, the U.S. Environmental Protection Agency, the Department of Health and Human Services, and the Consumer Product Safety Commission.

The Nanoscale Science, Engineering, and Technology Subcommittee of NNI formed the Nanotechnology Environmental and Health Implications (NEHI)

Working Group in 2005. The use of nanotechnology in the food safety industry is here—and if projections are correct, it will continue to grow rapidly over the next few years. Its potential for providing safer and more nutritious foods is important; however, there are still many uncertainties about the technology and its applications. We must make sure that the technology is safe and properly regulated to ensure maximum food safety and personal health protection. Exposure is a key element in risk assessment of nanostructures.

The environmental impacts of nanotechnology have become an increasingly active area of research. Until recently the potential negative impacts of nanomaterials on human health and the environment have been rather speculative and unsubstantiated. However, within the past number of years several studies have indicated that exposure to specific nanomaterials, *e.g.* nanoparticles, can lead to a gamut of adverse effects in humans and animals. This has made some people very concerned drawing specific parallels to past negative experiences with small particles.

Some types of nanoparticles are expected to be benign and are FDA approved and used for making paints and sunscreen lotion etc. However, there are also dangerous nanosized particles and chemicals that are known to accumulate in the food chain and have been known for many years.

In response to the above concerns, researchers have discovered a new bio-safe and bio-compatible route for the synthesis of magnesium oxide nanostructures. The route is based on simple reaction of water and magnesium powder. Since water is regarded as a benign solvent and non- toxic, the product (nanostructures) could be used safely for biomedical and other applications.

M.A. Shah of Taif University, Saudi Arabia and A.H Quarshi, of Toyma University, Japan both hailing from Indian part of Kashmir have developed a versatile approach for making magnesium oxide nanostructures that avoids organic solvents altogether. In their approach, magnesium powder is added to water, blasted with ultrasound for few minutes and then warmed at a temperature of 100 Celsius for 24 hours. X-ray and field emission electron microscopy studies reveal the product to be uniform nanoneedles of 40 to 60 nanometres.

We have found that we can use pure water in a simple and straightforward approach…suitable for large scale production. The proposed method is fast, economical, environmentally benign and free of pollution, which will make it suitable for large scale production. The health and environmental effects of common metals and materials are well-known.

The question for the researchers is whether the effects change when the metals and materials take the form of nanoparticles/nanorods – and whether these nanoparticles become more or less hazardous based on shape and size. "It's complicated," shah said. "In many cases, we lack basic knowledge of the properties and the behaviour of the particles themselves." There are already more than 400 companies worldwide that tap nanoparticles and other forms of nanotechnology, and regulatory agencies such as the Environmental Protection Agency, the Food and Drug Administration and the Occupational Health and Safety Administration are closely examining whether new regulations are needed to guard against potentially harmful but currently unknown effects.

It is worth mentioning that within a decade, nanotechnology could be used in nearly half of new products from computer devices to cancer like disease treatments, renewable energy sources, lightweight manufacturing components in cars and aeroplanes, agents for environmental remediation, water filters for removal of viruses, contaminants and salt for the entire world.

Such potential strides explain why nanotechnology is viewed as key to future economic growth and why technologically advanced countries are earnestly pursuing its development across the globe. Long-term developments are likely to occur in convergence with other emerging technologies, such as biotechnology and information technology, where nanotechnology will serve as an enabler of the new product.

It creates an opportunity to integrate education across physical and biological sciences, the social sciences and even humanities. Such integration is emblematic of new ways of thinking about the future and the workplace. The new technology will provide technological platform for industry, medicine and overall economy. Mad cow disease is a fatal neurodegenerative condition in cattle that is related to the human form of a disease that has caused the deaths of nearly 200 people worldwide.

Currently, testing for this disease in cattle is a lengthy process that only occasionally results in a correct diagnosis. With funding from USDA's Cooperative State Research, Education, and Extension Service (CSREES) National Research Initiative (NRI), scientists in New York created a new device that may provide a faster, easier, and more reliable way to test for mad cow disease, also known as bovine spongiform encephalopathy (BSE).

This new tool targets prions, which are the cause of BSE. Prions are abnormally structured proteins that convert normal proteins into an abnormal form. Prions are responsible for forms of the neurodegenerative diseases, such as BSE in cattle, scrapie in sheep, and Creutzfeldt-Jakob disease in humans. If often takes years before the symptoms arise that indicate the disease is present. The only test currently available for BSE involves multiple steps, requires sacrificing an animal host, and takes time. The process requires infecting an animal with a patient's blood. Then, after a several month incubation period, the animal is sacrificed and scientists look for prions during the animal's autopsy. This method produces the correct diagnosis only 31 per cent of the time.

A better method of prion detection is necessary to allay public fears, ensure the safety of the nation's food supply, and enhance international trade. Harold Craighead and colleagues at Cornell University have developed nanoscale resonators, which are tiny devices that function like tuning forks by changing pitch with increased mass. Craighead's group, in collaboration with Richard Montagna at Innovative Biotechnologies International, Inc., modeled the device after a similar idea used to detect bacterial pathogens.

When prions bind to the resonator's silicon sensor, it changes the vibrational resonant frequency of the device. In experimental trials, the sensor detected prions at concentrations as low as two nanograms per milliliter, the smallest levels measured to date. Currently, the resonator only detects prions in a saline solution. Efforts are now underway to use the resonator to detect prions in more complex solutions, such as blood.

"The real challenge is going to be to build an automated device that can take blood from a cow in the field and give a rapid response as to whether prions are present," Craighead said. "At the moment we only test cows when they fall over, but that is a late stage of the disease. It would be ideal to test cows a lot earlier. Resonators could be one path to doing this."

Scientists hope the new device will soon be used to detect prions in food items to ensure food safety and quality for the national food supply. CSREES funded this research project through the NRI Nanoscale Science and Engineering for Agriculture and Food Systems programme. Through federal funding and leadership for research, education and extension programmes, CSREES focuses on investing in science and solving critical issues impacting people's daily lives and the nation's future.

CHEMICAL HAZARDS ASSOCIATED WITH THE FOOD SUPPLY

Changes in our society and our food supply have raised new concerns about food chemical safety. Some of the changes that have raised new concerns about foodborne infectious disease are also affecting how government agencies carry out the more familiar task of protecting the food supply from toxic chemical agents.

Foods are themselves a complex collection of naturally occurring chemicals that have nutritive, organoleptic, and pharmacological functions and occasionally toxic effects. Naturally occurring toxicants probably present risks second only to those imposed by microorganisms. These natural toxicants include seafood toxins and foodborne mycotoxins.

Other chemicals are introduced into foods intentionally or unintentionally. Intentionally added substances include food and colour additives, flavours, enzymes, vitamins and minerals, and other ingredients that help to add value or characteristic properties or functions to a food. Because of broad public concern about synthetic chemicals, the toxicologic profiles of many of these synthetic components of food are much more complete than those of natural components of food. As a result, the risks posed by regulated food additives are generally better characterized than those of many naturally occurring substances. Unintentional additives can include environmental or industrial contaminants as well as some substances used in food production but not intended to be part of the food. The migration of food production substances into the final product is generally very low, but must be carefully regulated to ensure safety. Examples include sanitizers used to keep food production surfaces safe, packaging materials used to keep food safe and fresh, pesticides used on crops and drugs used in animals to mitigate damage, disease, microbial toxin production, and general food losses. Environmental or industrial contaminants are not sanctioned but have the potential to enter the food chain. Examples of chemical contamination incidents are methyl mercury in fish, mistaken mixing of polybrominated biphenyls (a fire retardant) into animal feed, and leakage of ammonia refrigerant into frozen foods.

Some people are sensitive or allergic to chemical constituents that are harmless to the rest of the population. These allergic reactions are estimated to occur in 1 to 2 per cent of adults and 5 to 8 per cent of children. Although serious reactions are rare, it is estimated that several dozen deaths occur each year because of allergic reactions to food. More than 160 foods and food-related substances have been identified as being able to cause allergic reactions. However, in the United States, more than 90 per cent of allergic reactions appear to be caused by just eight food types: peanuts, tree nuts (such as walnuts, pecans, almonds, hazel nuts), crustacea (shrimp, lobster, and other shellfish), eggs, milk, soy, fish, and wheat.

Consequently, the subject of chemical hazards in food is complex and includes consideration of potential risks that vary widely in scope and severity. Many chemical hazards associated with foods have been recognized only in the last century as advancements have been made in chemistry, toxicology, and risk assessment. Public concern over these hazards has grown in recent decades, in line with the increasing distrust of chemicals generally and of their use in the environment. Indeed, there have been episodes of chemical intoxication (usually arising out of accident or occupational exposure) with tragic consequences. Although most experts agree that the more serious hazards in the American food supply are not chemical but microbiological, public concern has demanded that proportionally more regulatory resources be applied to chemical hazards.

NEW FOOD COMPONENTS

The American public's growing interest in the relationship between diet and health has led to an increased demand for foods or food constituents that not only have nutritive value but also hold promise for prevention or even treatment of disease. These products have been referred to as dietary supplements, functional foods, pharma-foods or nutraceuticals. In a recent study, more than two-thirds of surveyed households reported use of a vitamin, mineral, or herbal supplement within the previous six-month period. The Dietary Supplement Health and Education Act of 1994 eased restrictions on certain statements of nutrition support made for supplements and exempted them from the safety approval requirement applicable to conventional food additives. That legal change helped spur the growing market for supplements of all types, including herbal products, and raises food safety concerns. Some supplements and herbal products on the market may pose a risk of adverse health effects because they are not required to meet specified safety standards before being sold. They may thus contain varying amounts or unknown or inadequately characterized ingredients that can have pharmacological activity that has not been adequately characterized.

Food processors are examining relatively new sources of ingredients for more conventional functional properties. For example, gums and fibres such as konjac flour, tara gum, inulin, or psyllium fibre may provide bulk and texture to foods, yet have had limited food use in the United States. Potential broad use of these ingredients in domestic foods raises important safety questions: are there significant intestinal effects such as blockage or reduced transit times? Are there impacts on vitamin uptake from the intestinal tract or on osmotic balance? What are the potential allergy risks from increased exposure to plant-sourced materials? These are a few of the safety questions that may be raised when a new ingredient source is considered.

Food processors are also utilizing macronutrient substitutes, such as non-nutritive sweeteners and fat replacers in many food products. These macronutrient substitutes have potential value by lowering calories, sugar, or fat in food products. Because dietary quantities of these substitutes could be substantially higher than the amounts of typical food additives, the assessment of their safety can be particularly important as well as particularly difficult.

New Food Technologies

Modification of plants or animals via genetic engineering can improve yields and increase resistance to pests. This new technology might offer improvements in food safety through increased resistance to molds that produce food mycotoxins or through lower levels of allergenic proteins, fatty acids, or other undesirable components of food.

However, there are important differences between countries in how food products from genetically modified organisms are regulated. Several products derived from genetic engineering have been declared safe by US regulators, but many European countries have either forbidden their sale or insisted on what marketers believe is disparaging labelling. Concerns about the safety of products from genetically engineered plants and animals are only partially resolved.

Food irradiation is not a new technology. Irradiation of fruits and vegetables decreases the risk of pathogens and extends shelf life. But as discussed, public concerns about the safety of irradiated food, fuelled in part by a lack of consumer education, limit its use in the United States. These concerns prompted the requirement that certain irradiated foods be labelled as such. The extent to which this labelling requirement has limited subsequent adoption of the process is unknown.

Re-emerging Toxic Agents

As the science of toxicology has progressed over the last 50 years, questions about the safety of chemicals in foods have become more sophisticated. When tougher safety standards are applied to new food chemicals or reapplied to old ones, new issues of toxicity emerge. The cycle of re-evaluating safety standards began in the years after World War II when advances in the understanding of the mechanisms of carcinogenesis coincided with the increased use of rodent bioassays. The newly focused attention to carcinogens that ensued led Congress to pass the 1958 "Delaney clause" proscription of carcinogens. The focus on carcinogens in food continues to consume substantial testing and regulatory resources. Further advances in biomedical research have raised new concerns and new standards of safety that now address teratogenicity, reproductive toxicity, mutagenicity, hormonal effects, and immunotoxicity.

The effects of food constituents on hormonal function is of concern as medical research has indicated the importance of certain hormones in regulating diseases such as breast cancer and osteoporosis. This phenomenon, sometimes referred to as endocrine disruption, led Congress in 1996 to mandate special regulatory attention to the issue and inspired the Environmental Protection Agency (EPA) to convene an expert panel, the Endocrine Disrupter Screening and Testing Advisory Committee.

That committee has attempted to define endocrine disrupters, determine analytical methods, and recommend how the disrupters should be regulated. Potential candidates for evaluation include constituents of some food packaging materials, pesticides, and natural food constituents, such as genestein in soy.

Some groups have recently asked whether chemical safety assessments provide adequate safety margins to protect children. That concern draws attention because new pesticide legislation directs EPA to consider the aggregate risks posed by any pest control agent; that is, exposures should be assessed for all potential sources of a chemical and for other chemicals with a similar mechanism of action. Another concern is the emergence of evidence of subtle developmental and behavioural effects of relatively low concentrations of chemicals (for example, lead).

In summary, chemical hazards, as well as foodborne pathogens, present new and changing challenges to the food safety system. Federal food safety efforts will need full integration and sufficient support to meet these challenges.

Physical Hazards

The foods we consume begin as raw agricultural commodities grown in open fields or waters or raised in a variety of production facilities, such as barns, coops, pens, and feedlots. Rocks, stones, metal, wood, glass, and other physical objects can become part of raw ingredients. Further contamination can occur in the transport, processing, or distribution of foods because of equipment failure, accident, or negligence.

Foreign physical materials in foods can cause serious harm to consumers. Protective devices that remove or prevent physical hazards include metal detectors, magnets, sieves, traps, scalpers, and screens. Other effective means of protection are production plant policies against the use of glass, wood, or non-ferrous metal where possible; employee training; quality audits of ingredient suppliers; and sensory tests.

Federal agencies have established "defect action levels" for natural or unavoidable defects that do not affect human health, such as stems or pits in fruits and vegetables, bone in mechanically deboned meats, and microscopic insect fragments, which are primarily of aesthetic concern and do not present a safety hazard.

The Changing Nature of Food Hazards

- Changes in the risk of infectious foodborne disease are due primarily to:
- Changes in diet,
- Increasing use of commercial food service and in food eaten or prepared away from home,
- New methods of producing and distributing food,
- New or re-emerging infectious foodborne agents, and
- The growing number of people at high risk for foodborne illnesses due to:
  - Increasing number of elderly, and
  - Increasing number of people with depressed immunity or resistance to infection.
- Changes in chemical hazards associated with the food supply must be monitored and evaluated; these include:
- Increased use of dietary supplements and herbal products without requirements to meet specified safety standards;
- New food components that mimic attributes of traditional food components;
- Introduction of new food technologies and processes;
- Changes in presence of food toxins and additives, including unintentional food additives; and
- Presence of physical hazards associated with new technologies or sources of foods.

CLIMATE CHANGE AND FOOD SECURITY

IMPACTS ON FOOD PRODUCTION AND AVAILABILITY

Climate change affects agriculture and food production in complex ways. It affects food production directly through changes in agro-ecological conditions and indirectly by affecting growth and distribution of incomes, and thus demand for agricultural produce. Impacts have been quantified in numerous studies and under various sets of assumptions. A selection of these results is presented in *Quantifying the Impacts on Food Security*. Here it is useful to summarize the main alterations in the agro-ecological environment that are associated with climate change.

Changes in temperature and precipitation associated with continued emissions of greenhouse gases will bring changes in land suitability and crop yields. In particular, the Intergovernmental Panel on Climate Change (IPCC) considers four families of socio-economic development and associated emission scenarios, known as Special Report on Emissions Scenarios (SRES) A2, B2, A1, and B1. Of relevance to this review, of the SRES scenarios, A1, the "business-as-usual scenario," corresponds to the highest emissions, and B1 corresponds to the lowest.

The other scenarios are intermediate between these two. Importantly for agriculture and world food supply, SRES A2 assumes the highest projected population growth of the four (United Nations high projection) and is thus associated to the highest food demand. Depending on the SRES emission scenario and climate models considered, global mean surface temperature is projected to rise in a range from 1.8°C (with a range from 1.1°C to 2.9°C for SRES B1) to 4.0°C (with a range from 2.4°C to 6.4°C for A1) by 2100. In temperate latitudes, higher temperatures are expected to bring predominantly benefits to agriculture: the areas potentially suitable for cropping will expand, the length of the growing period will increase, and crop yields may rise.

A moderate incremental warming in some humid and temperate grasslands may increase pasture productivity and reduce the need for housing and for compound feed. These gains have to be set against an increased frequency of extreme events, for instance, heat waves and droughts in the Mediterranean region or increased heavy precipitation events and flooding in temperate regions, including the possibility of increased coastal storms; they also have to be set against the fact that semiarid and arid pastures are likely to see reduced livestock productivity and increased livestock mortality.

In drier areas, climate models predict increased evapotranspiration and lower soil moisture levels. As a result, some cultivated areas may become unsuitable for cropping and some tropical grassland may become increasingly arid. Temperature rise will also expand the range of many agricultural pests and increase the ability of pest populations to survive the winter and attack spring crops.

Another important change for agriculture is the increase in atmospheric carbon dioxide (CO_2) concentrations. Depending on the SRES emission scenario, the atmospheric CO_2 concentration is projected to increase from H"379 ppm today to >550 ppm by 2100 in SRES B1 to >800 ppm in SRES A1FI. Higher CO_2 concentrations will have a positive effect on many crops, enhancing biomass accumulation and final yield. However, the magnitude of this effect is less clear, with important differences depending on management type (*e.g.*, irrigation and fertilization regimes) and crop type. Experimental yield response to elevated CO_2 show that under optimal growth conditions, crop yields increase at 550 ppm CO_2 in the range of 10 per cent to 20 per cent for C_3 crops (such as wheat, rice, and soybean), and only 0–10 per cent for C_4 crops such as maize and sorghum. Yet the nutritional quality of agricultural produce may not increase in line with higher yields. Some cereal and forage crops, for example, show lower protein concentrations under elevated CO_2 conditions.

Finally, a number of recent studies have estimated the likely changes in land suitability, potential yields, and agricultural production on the current suite of crops and cultivars available today. Therefore, these estimates implicitly include adaptation using available management techniques and crops, but excluding new cultivars from breeding or biotechnology. These studies are in essence based on the FAO/International Institute for Applied Systems Analysis (IIASA) agro-ecological zone (AEZ) methodology. For instance, pioneering work in ref. 9 suggested that total land and total prime land would remain virtually unchanged at the current levels of 2,600 million and 2,000 million hectares (ha), respectively. The same study also showed, however, more pronounced regional shifts, with a considerable increase in suitable cropland at higher latitudes (developed countries +160 million ha) and a corresponding decline of potential cropland at lower latitudes (developing countries "110 million ha). An even more pronounced shift within the quality of cropland is predicted in developing countries. The net decline of 110 million ha is the result of a massive decline in agricultural prime land of H"135 million ha, which is offset by an increase in moderately suitable land of >20 million ha. This quality shift is also reflected in the shift in land suitable for multiple cropping. In sub-Saharan Africa alone, land for double cropping would decline by between 10 million and 20 million ha, and land suitable for triple copping would decline by 5 million to 10 million ha. At a regional level, similar approaches indicate that under climate change, the biggest losses in suitable cropland are likely to be in Africa, whereas the largest expansion of suitable cropland is in the Russian Federation and Central Asia.

Impacts on the Stability of Food Supplies

Global and regional weather conditions are also expected to become more variable than at present, with increases in the frequency and severity of extreme events such as cyclones, floods, hailstorms, and droughts. By bringing greater fluctuations in crop yields and local food supplies and higher risks of landslides and erosion damage, they can adversely affect the stability of food supplies and thus food security.

Neither climate change nor short-term climate variability and associated adaptation are new phenomena in agriculture, of course. As shown, for instance, in ref. 9, some important agricultural areas of the world like the Midwest of the United States, the northeast of Argentina, southern Africa, or southeast Australia have traditionally experienced higher climate variability than other regions such as central Africa or Europe. They also show that the extent of short-term fluctuations has changed over longer periods of time. In developed countries, for instance, short-term climate variability increased from 1931 to 1960 as compared with 1901 to 1930, but decreased strongly in the period from 1961 to 1990. What is new, however, is the fact that the areas subject to high climate variability are likely to expand, whereas the extent of short-term climate variability is likely to increase across all regions. Furthermore, the rates and levels of projected warming may exceed in some regions the historical experience.

If climate fluctuations become more pronounced and more widespread, droughts and floods, the dominant causes of short-term fluctuations in food production in semiarid and subhumid areas, will become more severe and more frequent. In semiarid areas, droughts can dramatically reduce crop yields and livestock numbers and productivity. Again, most of this land is in sub-Saharan Africa and parts of South Asia, meaning that the poorest regions with the highest level of chronic undernourishment will also be exposed to the highest degree of instability in food production. How strongly these impacts will be felt will crucially depend on whether such fluctuations can be countered by investments in irrigation, better storage facilities, or higher food imports. In addition, a policy environment that fosters freer trade and promotes investments in transportation, communications, and irrigation infrastructure can help address these challenges early on.

Impacts of Climate Change on Food Utilization

Climate change will also affect the ability of individuals to use food effectively by altering the conditions for food safety and changing the disease pressure from vector, water, and food-borne diseases. The IPPC Working Group II provides a detailed account of the health impacts of climate change of its fourth assessment report. It examines how the various forms of diseases, including vector-borne diseases such as malaria, are likely to spread or recede with climate change. This chapter focuses on a narrow selection of diseases that affect food safety directly, *i.e.*, food and water-borne diseases.

The main concern about climate change and food security is that changing climatic conditions can initiate a vicious circle where infectious disease causes or compounds hunger, which, in turn, makes the affected populations more susceptible to infectious disease. The result can be a substantial decline in labour productivity and an increase in poverty and even mortality. Essentially all manifestations of climate change, be they drought, higher temperatures, or heavy rainfalls have an impact on the disease pressure, and there is growing evidence that these changes affect food safety and food security.

The recent IPCC report also emphasizes that increases in daily temperatures will increase the frequency of food poisoning, particularly in temperate regions. Warmer seas may contribute to increased cases of human shellfish and reef-fish poisoning (ciguatera) in tropical regions and a poleward expansion of the disease. However, there is little new evidence that climate change significantly alters the prevalence of these diseases. Several studies have confirmed and quantified the effects of temperature on common forms of food poisoning, such as salmonellosis. These studies show an approximately linear increase in reported cases for each degree increase in weekly temperature. Moreover, there is evidence that temperature variability affects the incidence of diarrhoeal disease. A number of studies found that rising temperatures were strongly associated with increased episodes of diarrhoeal disease in adults and children. These findings have been corroborated by analyses based on monthly temperature observations. Several studies report a strong correlation between monthly temperature and diarrhoeal episodes on the Pacific Islands, Australia, and Israel.

Extreme rainfall events can increase the risk of outbreaks of water-borne diseases particularly where traditional water management systems are insufficient to handle the new extremes. Likewise, the impacts of flooding will be felt most strongly in environmentally degraded areas, and where basic public infrastructure, including sanitation and hygiene, is lacking. This will raise the number of people exposed to water-borne diseases (*e.g.*, cholera) and thus lower their capacity to effectively use food.

MODERN FOOD SUPPLY AND SAFETY

Around the world there is an increase in interest in our food, health and environment. Where are products produced, how, why, by whom, how far have they travelled, how long have they been stored etc. The organic and local food movements have emerged as an intuitive and practical response to the increasing use of chemicals in food production, and to the growing alienation of industrial agribusiness from holistic agricultural systems.

People have chosen to eat organic foods because they care about the health of their families and the health of the environment. Organic agriculture also enables people to support integrated, environmentally friendly agriculture, and appropriate technology, rather than chemical-intensive factory farming. Support for organics has also grown as a direct response to biotechnology giants' efforts to genetically engineer our food crops.

Farmers and food buyers around the world were, and continue to be, enraged by the introduction of genetically engineered food crops. For many, the inevitable conclusion was that whereas the biotechnology companies stood to benefit from the entry of genetically engineered foods into the food chain, consumers, farmers and the environment shouldered all the risks. Now, nanotechnology introduces a new wave of assaults on our foods.

Nanotechnology is the high technology, atomically processed antithesis to organic agriculture, which values the natural health-giving properties of fresh, unprocessed wholefoods. It further transforms the farm into an automated extension of the high technology factory production line, using patented products that will inevitably concentrate corporate control. It also introduces serious new risks for human health and the environment. Regularized and improved food supply has been a consistent factor throughout history in improving human health and longevity.

In more recent times, improved food safety has added another dimension to human health. The evidence on the safety of our modern food supply is quite contrary to the strongly held conventional wisdom. Humans have had to deal with food contamination throughout history. Human foodstuffs have been the carriers of botulism, ergot, and the aflatoxins from the fungus *Aspergillus flavus,* which has caused mass illness, blindness, and largescale death. Even uncontaminated natural foods contain substances that would be considered a threat to human life if they were used as a food additive. In the United States, during the early twentieth century, contaminated food, milk, and water caused many foodborne infections, including typhoid fever, tuberculosis, botulism, and scarlet fever.

Once the sources and characteristics of foodborne diseases were identified, there were technologies that could help to control them before vaccines or antibiotics: "handwashing, sanitation, refrigeration, pasteurization, and pesticide application." Healthier animal care, feeding, and processing also improved food supply safety.

PLANT-PRODUCED CHEMICALS

The buzz word "chemicals," used by some to label and condemn food additives, could appropriately be used to describe what the plants themselves manufacture. Plants of all kinds, and most other life forms, are chemical factories, manufacturing a variety of chemicals, some of which have very active properties.

Humans have extracted some of these chemicals for medicinal uses, some as poisons. Given the truism that dose makes the poison, some chemicals from plants that were once used as poisons are now used for medical purposes. In other words, chemicals from plants are not inherently good or bad; goodness or badness is determined by a variety of factors and the uses to which they are put.

One of the reasons for maintaining as much biodiversity as possible is that the earth's plants produce an array of active chemicals, the vast majority of which have not been screened for possible human use. According to Katharine Milton, plants are unable to run from hungry predators and so have developed a variety of defences to avoid the loss of edible components.

"These protections include a vast array of chemicals known as secondary compounds (such as tannins, alkaloids and terpenoids). At best these chemicals taste awful; at worst, they are lethal". Leopold and Ardrey add that these plant secondary substances are "chemicals that do not participate in the basic metabolism of the plant.

Among these are many chemicals that serve to repel or discourage the use of the plant by insects, microorganisms, nematodes, grazing animals, and man." These plant toxins apparently are often used to ward off fungi. "While antifungal chemicals in nightshade were distasteful ... there are probably protective compounds in other wild fruits that prevent rotting without spoiling taste, compounds that should be of particular interest to plant breeders".

Many plants produce "phytoestrogens," the so-called estrogen disrupters, which might have the effect of reducing the fertility of their predators, the animals that graze or otherwise consume them. "Essentially, they give their predators an oral contraceptive". Eating fresh, uncooked produce may have many taste and health benefits, but it can also be dangerous unless proper precautions are taken. As epidemiologist Michael T. Osterholmm stated: "The hearthealthy diet may be kind to our cardiovascular system but it is hell on the digestive tract". Rachel Carson, whose book *Silent Spring* is seen as the inspiration of the modern environmental movement, recognized the role of natural toxins—systemic pesticides—in plants.

She maintained that the perception of nature as an "enchanted forest" was a "fairy tale". Even some of the weeds associated with cultivars and often eaten with them are endowed with toxins and other substances deleterious to human health. Garn notes, "Weeds growing along with cultivars often included plants with interesting alkaloids, many of them in the tomato family that includes *Datura, Belladonna,* and the like".

TOXINS IN PLANT EVOLUTION

Through time, plants evolved for their own survival, not to serve human needs.

"Plants in nature synthesize toxic chemicals in large amounts, apparently as a primary defence against the hordes of bacterial, fungal, and insects and other animal predators *plants in the human diet are no exception.*" The evolutionary process involved in domestication changed the survival strategy for these plants, as they now depended upon the continued cultivation to survive.

The toxicity of many plants was reduced compared to their wild progenitors, but in most cases was not completely bred out of plants. In fact, currently there is considerable effort to breed plants with greater disease resistance so as to reduce the need for pesticides, in some instances by increasing the production of these toxins; in other instances, not.

GREEN REVOLUTION PROS AND CONS

It is hard to imagine anyone opposing the green revolution. The increase in yields from the green revolution in rice alone has produced enough to feed one billion people.

Growth in yields accounts for 92 per cent of the increase in world cereal production since 1960. The world average for grain yields per hectare rose from 1.1 tons in 1950 to 2.9 tons in 1992. Without the increase in yields, it would have been necessary to bring another 3.6 billion hectares of land under cultivation, which would have almost doubled world cropland from "34 per cent of the earth's surface to 61 per cent". The green revolution was in many ways a grain revolution— wheat, rice, and maize—but it also was a research revolution that facilitated an extraordinary and sustained general expansion in food production.

Bender and Smith note, "Between 1961 and 1994, the number of daily food calories per capita rose from about 1,900 to 2,600 in developing countries, while their populations nearly doubled from 2.2 billion to more than 4.3 billion." Globally in the same period, "average daily per capita food supplies increased more than 20 per cent."

The increase in available calories per capita for the developing countries rose 50 per cent from 1948–1952 to 1994–1996. A century-long trend of falling real food prices continued during the period 1950 to 1992 as international food commodity prices dropped 78 per cent in constant 1990 prices. D. Gale Johnson sums up these achievements in agriculture: "People today have more adequate nutrition than ever before and acquire that nutrition at the lowest cost in all human history."

In the 1990s, while population continued to increase, the absolute number of malnourished people in the world declined by forty million, which is a continuation of a trend of the last decades. However, this was considered a "failure" by the UN World Food Programme because it fell short of the reductions that were thought to be possible. To the extent that it is a failure, it is a failure of institutions, not of the technology that is producing the food that could make the programme a success. So, there are arguments critical of the green revolution.

The first argument is that it was a failure. Data such as that just cited above have demolished that argument, although a few still assert it, unaware of the change in the antitechnology party line. Next, we are told that the new varieties may have increased yields, but they led to a worsening of income distribution and the condition of the very poor. As early as the 1970s, studies emerged showing that the poor consumers and farmers benefited proportionately more than other groups as part of the overall improvement in human wellbeing. Some argued that tenant farmers in the Punjab in the north of India were displaced from their land due to high-yielding varieties of wheat.

It turns out that they returned to the land as farm laborers and earned more than they did as tenant farmers. Similarly, in southern India, between 1973 and 1994, the average real income of small farmers increased by 90 per cent, and that of the landless— among the poorest in the farm community—by 125 per cent. Further, caloric intakes for small farmers and the landless rose 58 to 81 per cent, and protein intakes rose 103 to 115 per cent. In addition, greater farm income meant more income and employment in the local villages that supplied the farm with goods and services now affordable to those who worked the land.

The green revolution in Indian agriculture is credited with being a major force in poverty reduction in India.

From independence to 1970, the per centage of the population of India living below the absolute poverty line remained relatively steady at 55 per cent. In over two decades since the introduction of the green revolution technologies, the per centage of the population of India living in absolute poverty has fallen to 35 per cent. Throughout most of East and Southeast Asia, the early phases of development involved an emphasis on agriculture and food production. The green revolution technologies in agriculture have played an important role in the dramatic reductions in poverty rates throughout the region, such as in Indonesia from 58 per cent in 1970 to 8 per cent in 1993, and in China from 270 million people in poverty in 1978 to 65 million in 1999.

Yao estimates the reduction in rural poverty in China to be from between 596 to 790 million people (75.5 per cent–100 per cent) in 1978 to 57 to 114 million (6.7 per cent–13.2 per cent) in 1996 (447). By either measurement, the magnitude of the reduction in poverty in China is extraordinary and without precedent in human history.

In the time period 1969–1971 to 1990–1992, global population was increasing about 45 per cent while the absolute number of people suffering from chronic undernourishment in developing nations decreased from 917 million or 35 per cent of their population to 839 million or 21 per cent.

From 1990–1992 to 1995–1997, the absolute number of undernourished people in the developing world continued to decline, falling to 790 million. The prevalent high number should not be minimized, nor should we be complacent, but neither should the progress achieved be ignored. For we can finish the job of eliminating undernourishment by recognizing the forces that have brought us this far, thereby gaining some understanding of what still needs to be done. Unfortunately, many wish to deny the progress that has been made, offering antitechnology solutions, which would reverse these gains.

ALTERNATIVE MEDICINE

Some of the most dangerous ingested products legally sold are found in health food stores. Herbal dietary supplements (including exotic oriental herbs) and amino acids pills have been causing health problems for some of their users.

A chemical additive may or may not be inactive, but the herbs and amino acids are taken precisely because they are active and will alter a bodily function. Most of them are untested and unregulated. To the extent that testing is done, there is a critical difference between the approach to herbal medicine and that of pharmaceuticals.

Drugs are not permitted to be used until proven to be safe and effective, while the philosophy and practice of alternative medicine has operated on the "presumption that treatments are safe and effective unless proven otherwise". Alternative medicine is big business.

One study estimated that in 1990, "Americans poured $13.7 billion into alternative remedies, and made more visits to alternative healers than to primary-care doctors". An editorial in the *New England Journal of Medicine* clearly defines the difficulties with alternative medicine: "What most sets alternative medicine apart is that it has not been scientifically tested and its advocates largely deny the need for such testing". Lacking is "the marshaling of rigorous evidence of safety and efficacy" necessary for approval of drugs and for publication in the "best peer-reviewed medical journals".

One respected study has found that Americans are spending $27 billion a year on alternative treatments, which is close to their outof-pocket expenditures on modern medical treatment. Some herbs and treatments were found to be beneficial, but some were not. Investors have found the stocks of the firms manufacturing these dietary supplements and alternative medications to be attractive, indicating a belief that consumers will buy them, whether proved to be safe and effective or harmful.

Even for some herbs that have been used in low dosage for centuries, there are no studies on the impact of the mega doses that are often taken. The herbs and alternative medicine advocates have succeeded in blurring the line between food and drugs. Herbal extracts sold as food supplements are being added to packaged, canned, and frozen food products—called functional-foods—with labels that imply health benefits and have sufficient growth potential to have attracted some of the major food producers. Some of the early advocates of alternative foods and medications, like Rodale Inc. and its publication, *Prevention,* recognize the danger in current patterns of use of herbs and dietary supplements.

Prevention's national survey on the self-care movement reveals "158 million consumers use dietary supplements for their health and spend $8.5 billion each year." It reported that "widespread use of dietary supplements may cause public health problems". There were risks mixing dietary supplements with prescription drugs and treating oneself without a doctor's supervision. The survey estimated that "11.9 million customers have experienced adverse reactions from using herbal remedies and 6.5 million have had problems of this kind when using specialty supplements". The problems with various herbs are legion, including lack of long-term studies and of any kind of control of dosage intake.

Since many who use alternative medical treatment are also taking prescription drugs under the care of medical doctors, there may be substantial health risks, including death, in combining these with herbs and other alternative medications and stopping the use of some herbs after one had been taking them. Serious problems have emerged from ingesting some of the most popular herbs, such as *Echinacea purpurea,* St. John's Wort, or Ginkgo Biloba.

Probably the most serious harm can come from the interaction between herbs and prescription medicines. The increasing provision of these herbs by parents to their children is evident in the statistics of the American Association of Poison Control Centers.

One herbal remedy sometimes called ephedra, sometimes called ma huang, sometimes called pseudoephedrine, contains the chemical ephedrine, "a potent drug whether it is synthetically manufactured or produced naturally from a plant". Dr. Robin R. Caldwell says the drug is "chemically almost identical to amphetamine". Not only is ma huang sold in health food stores, it is a listed ingredient in many products sold at truck stops and elsewhere to help drivers stay awake.

Pseudoephedrine is in various cold medicines that carry label warnings not to be used by anyone with high blood pressure, heart trouble, or liver problems. Ma huang is also used in diet and pep pills. Because ephedra is classified as a dietary supplement under the 1994 Dietary Supplement Health and Education Act, the FDA can restrict its sale "only if it receives well-documented reports of health problems associated with it."

Brody notes, "The agency took four years, and more than 100 reports of life-threatening symptoms and 38 deaths, to act against ma huang, a stimulant that can prove disastrous to people with heart problems." Warnings have also been posted for Chinese herbal products containing aristolochia acid, which has been shown to cause cancer and kidney failure. Nearly one hundred women with kidney damage from Chinese herbs at a Belgian slimming clinic were studied, and several were found "in dire need of kidney transplants" and many of them were developing cancers in the kidney and bladder, years after taking the drug". Those opposing any regulation of dietary supplements are involved in a contradiction that doesn't seem to bother them.

These herbal and other dietary supplements are taken because they have some curative effect from an active ingredient or agent. Stated simply, "if it's strong enough to work, it's strong enough to hurt".

TECHNOLOGY AND LIFE EXPECTANCY

The presumed superiority of all things natural in food and other of life's necessities is the other side of the belief that the food supply in affluent societies is contaminated and is part of the larger, firmly held conviction that modern life is unsafe. One wonders, if contemporary life is so polluted, why are we living so long? Riley has argued that since we live longer and are older, we have more time at risk to ill health. In fact, morbidity has fallen, along with mortality. We live longer, healthier lives.

Among older Americans whom Riley deems to "have more time at risk to ill health," there has been a dramatic decline in chronic disability, according to the National Long Term Care Survey, as reported in a 1997 study by the National Academy of Sciences. "In the last decade alone, senior citizens have experienced a 25 per cent decline in the number of days of restricted activity due to illness". Improvement in the quality of life continues through the last year of life, even as we live longer. In the workplace where safety has been largely ignored by environmentalists, the Centers for Disease Control and Prevention has found that "work-related deaths declined from 61 per 100,000 workers in 1913 to four per 100,000 workers in 1997" with a 50 per cent decline in the risk of a fatal or disabling accident from 1970 to 1990.

Says Mays, "An estimated 23,000 people died from work-related injuries in 1913 compared with 5,100 in 1997, according to the National Safety Council and the U.S. Bureau of Labour Statistics. During this same period, the workforce grew from 38 million to about 130 million." Contrary to Riley, one might argue that the factors contributing to longer life are part of the same cluster of factors that reduce morbidity, including chronic illness. Fogel and Costa argue cogently, and with massive data, that the epidemiology of chronic disease is not separate from that for contagious disease. Beginning in utero or infancy, inadequate nutrition can lead to a vast array (far too many to list here) of deleterious conditions that make the offspring more susceptible to contagious diseases, to chronic illness later in life, and to shorter life expectancy.

ADVANCES IN CHILDBEARING

Giving birth to a child is dangerous to child and mother. One demographic study for the Plymouth Colony in the seventeenth century concluded that one in every thirty childbirths resulted in maternal mortality, which was about 20 per cent of all female deaths.

Add in the 25 per cent estimated death rate for infants and children (for both sexes), and the conclusion is that close to half—about 45 per cent assuming roughly equal infant and child death rates for the sexes—of all females died before completing fertility. This was in a society where, the author found, the "records suggest a standard of life and health that would compare favorably with that of any preindustrial society today".

Edward Shorter, in his controversial writing on the history of women's bodies, argued that virtually all women had some lifelong debilitating health problem resulting from giving birth to children. Given the high rates of maternal mortality, it is not unreasonable to assume that the debilitating injury rate was a multiple of at least five, putting the injury rate close to 100 per cent. Recent studies would seem to indicate that giving birth to a child, even under favorable conditions, limits a woman's ability to make the required "investments in somatic maintenance" for longevity and thereby shortens her life expectancy.

In earlier times, when nutrition levels were very much lower, and the disease experiences (both previous and current infections) of the expectant mother were much higher, it is very likely that there was organ damage to mother and child and that the mother's ability to make the "investments in somatic maintenance" for longevity were greatly diminished. If having children, even under the most favorable conditions, is harmful to the mother, then an additional benefit of lower infant mortality is that it allows a woman to achieve her desired family size at a much lower health cost to her.

This clearly reinforces the thesis of Fogel and Costa on the importance of adequate nutrition and protection from disease for the long-term health of both the mother and the child. Fogel and Costa call their theory "technophysio evolution. " They say, "The theory of technophysio evolution rests on the proposition that during the last 300 years, particularly during the last century, humans have gained an unprecedented degree of control over their environment".

This control sets humans not only "apart from all other species, but also from all previous generations of *Homo sapiens.*" They add, "This new degree of control has enabled *Homo sapiens* to increase its average body size by over 50 per cent, to increase its average longevity by more than 100 per cent, and to improve greatly the robustness and capacity of vital organs". The data on the decline in maternal mortality in developed countries (and to some extent for developing countries) in the twentieth century are every bit as dramatic as the declines in infant mortality and increase in life expectancy during this same time period.

The 1998 World Health Organization report confirms that the global trend towards longer and healthier lives continues, though concern is raised for the gap in life expectancy and good health between the rich and the very poor. WHO is now talking about "health expectancy" being as important an indicator of human well-being as life expectancy.

Herman says, "Life expectancy...in rich countries and relatively poor ones, has catapulted to marks unthinkable a century ago. In the industrial world 25 years in life expectancy have been gained in the 20th century, more than twice the gain in all previous human history."

LIFE EXPECTANCY

Many of the news articles on the global increase in life expectancy discuss the problems created by this increase and the social and economic changes necessary to accommodate it. The age at which people retire, the funding of retirement and medical care systems, and the need for education and training for second careers are all issues (or problems), among many others, that will need to be addressed.

In many cases they are already issues under serious discussion in economically advanced countries. The character of these systems will differ in the world in 2050, when one-fifth of the global population will be over sixty-five, from those few that existed at the beginning of the twentieth century when the first social security systems began and less than 1 per cent of the world's population was over sixty-five. Though the numbers differ from country to country, the magnitude of the change for most individual countries will be roughly comparable. Technological progress does not offer the promise of a problem-free utopia, but it does offer a never-ending stream of challenges with the ever-present possibility of continuing human betterment.

WATER AND PUBLIC HEALTH

Humanity did not regularly drink hygienically clean water until the advent of purification processes, which included adding chemicals, in the early twentieth century.

The late eighteenth and early nineteenth centuries brought developments in basic science, which allowed us to understand respiration and statistics that laid the foundations for quantitative medicine. Lavoisier's work led to the "secularization and demystification of water by analyzing it and showing that it could be broken down into hydrogen and oxygen".

In the middle of the nineteenth century, the physician John Snow identified the water from one well as the source for an outbreak of cholera, followed by Louis Pasteur's recognition of the microbial origin of many diseases, some of which were waterborne. Previously, water could be visually clean or ritually clean, but with this new knowledge and chemical intervention, we could have hygienically clean water. In short, "water became an industrial product". Chlorination of water in the United States began in the early part of the twentieth century and very quickly "produced dramatic reductions in morbidity and mortality associated with waterborne disease, such as typhoid, cholera, amoebic dysentery, bacterial gastroenteritis, and giardiasis."

Chlorinating contaminated drinking water stopped a potential typhoid epidemic in Chicago in 1908. The introduction of drinking water disinfection in the United States is credited with reducing the incidence of cholera by 90 per cent, typhoid and leptospirosis by 80 per cent, and amoebic dysentery by 50 per cent. A chlorinated lime solution was used as an antiseptic in hospitals in the nineteenth century. Prior to chlorination of water, diarrhea and enteritis were the third leading cause of death in the United States. Despite the fact that millions of lives have been saved by the use of chlorine for disinfection of water, there are some who would ban its further use, even though the evidence for its danger is meagre. Chlorination of water is one of the triumphs of the twentieth century, but many people in the world have yet to experience its benefit.

The best estimates are that 80 per cent of all diseases and more than one-third of deaths in developing countries are caused by drinking or cooking with contaminated water. About twenty-five thousand children die each day from waterborne diseases. The concerns expressed about the use of chlorine are for the chlorine compounds that emerge in combination with other chemicals in the water. For any other chemical used instead of chlorine, we would have to be concerned with its toxicity *and* with the toxicity of every possible compound that it might form in the effluent from our water system.

The environmentalist precautionary principle would simply rule against any substitute, particularly when set against the nearly century-long safe, effective use of chlorine. Because of U.S. risk assessments claiming that chlorine is a carcinogen, Peru stopped chlorinating its drinking water in 1991. What followed was the "largest outbreak of cholera in recent history, which killed nearly 7,000 people and afflicted over 800,000 more".

Regulators, "influenced by the public's present tendency towards chemophobia," at times fail to weigh risks and benefits adequately. An outbreak of *E. coli* O157:H7 occurred in one community in the United States where the spring water used in the public water system was not chlorinated. In response to the outbreak of *E. coli* O157:H7, the community began chlorinating its water supply. Even with seemingly adequate chlorine levels, an indoor swimming pool complex with "extensive water spray features disseminated pathogenic bioaerosols," causing serious lung infections among its young, healthy lifeguards.

Among corrective steps taken was the "provision of free residual chlorine in pool water from chlorine gas". In a study in Britain covering the years 1937 to 1986, "defective chlorination" was blamed in eight out of ten outbreaks of disease from public water supplies and in all thirteen outbreaks in private supplies. Costs of not using a technology are often not considered. The cure for not using a technology is to use more technology and use it more intelligently.

BOTTLED WATER AND TAP WATER

In Houston, people are buying bottled water, which is from the same source—the municipal water supply—that they can draw safely from their own tap more conveniently, at far less cost, and without a plastic bottle that will eventually be discarded and add to environmental pollution.

The municipal water supply as the source for the water is clearly printed on the labels of the bottles. "In fact, 25 per cent of all bottled water is tap water drawn from city water treatment plants, according to Bob Brady of the International Bottled Water Association," though the bottler may do things to the water after it is drawn. An estimate by the American Waterworks Association finds that as many as 50 per cent of bottled water manufacturers in the United States obtain their water from municipal water departments.

"This is the stupid thing. They buy water from the municipality for about 40 cents for 1,000 litres, they filter it two or three times, then they sell it at 2,000 times that price". Much of the bottled water that is not from the local municipal water supply is drawn from aquifers in rural areas. In the United States, where the legal principle of "right of capture" prevails, a major bottling company can come in, drill a few wells, and draw vast quantities of water.

In Texas, there have been charges (and denials) that companies are drawing down the aquifers, causing the existing wells of local inhabitants to go dry. The most important factor behind the extraordinarily rapid growth of the bottled water industry is the public perception that bottled water is cleaner and safer and tastes better than tap water. A study of the bacterial content of bottled water by the University of Mississippi questioned whether it is better than most municipal tap water.

FOODBORNE DISEASES FOR HUMANS

Foodborne diseases have always been a problem for humans, even for hunting and gathering peoples, says Cohen. Diseases might be caught from wild animals, including rabies or tularemia, which may have been "a significant cause of sickness and death among American Indian populations who regularly handled game and fur bearing animals."

Cohen adds, "Handling wild animals or their remains can also result in infection with such other diseases as toxoplasmosis, hemorrhagic fevers, leptospirosis, brucellosis, anthrax, salmonellosis people can encounter a variety of highly lethal anaerobic bacteria, including the agents of gangrene, botulism, and tetanus, if they expose themselves to the intestines of animals while butchering a kill". Add in trichina worms (trichinosis) and staphylococcal infections, and it is clear that food harvested in the wild is not necessarily clean, wholesome, and natural.

Drinking water in the wild can cause "beaver belly" (giardiasis) or be a source of "microorganisms derived from moose, ducks, and geese as well". Water in the wild today, including snow collected from the ground and melted, can be a source of a variety of parasites, including giardia and cryptosporidium. If we think that we eat contaminated food today, and that we ate clean food in earlier times, think again. Some foods require cooking or other forms of preparation to be digestible by humans, such as soaking in water, having husks or shells or other parts removed, or milling and reducing particle size to facilitate digestion. Even today, with small-scale technologies, grain milling is incomplete, leaving broken pieces of husk, causing stomach lesions and digestive problems.

In the past (and in many less-developed areas today), "harvesting and milling techniques resulted in a considerable accretion of grit and little stones, as we see in the worn teeth of our colonial ancestors, Bronze-age and Ironage Europeans, and American Indians." With storage, "weevils were often a regular part of the daily bread". On the bright side, food contaminants, such as weevils, were sometimes important sources of protein in otherwise protein-deficient diets.

HEALTH FOOD FADS

Health food faddists who look with horror on the minutest trace of any chemical (read manufactured chemical) nevertheless will ingest mega doses of natural products (such as herbs) that contain substances known to be toxins or untested compounds or have active products such as amino acids that have not been tested for safety at the dosage in which they are being taken. To call free-standing amino acids "natural" is a travesty, since they are not naturally present in food in significant quantity nor can they be "extracted from food without first chemically breaking food protein in the laboratory into its constituent amino acids".

It is ludicrous to label a bottle of amino acids "natural," and it shows how meaningless the term has become and how useless it is for formulating public policy. The amino acid pills are a synthetic product, produced in large vats in chemical factories using microbial fermentation, with a feedstock for the microorganism that is so replete with contaminants that it has to be further processed through a carbon filter and by reverse osmosis to produce what rarely occurs in nature, a free-standing amino acid. We get almost all of our amino acids by ingesting complex proteins. In 1989, L-tryptophan amino acid pills were identified as a cause of a rare muscle disorder, eosinophilia-myalgia syndrome, which eventually resulted in thirty-seven deaths and at least, 1,500 illnesses.

Activists blamed it on the fact that the manufacturer used a bacteria, *Bacillus myloliquefaciens,* that had been genetically engineered, and they continue to argue this to the present, despite the fact that the overwhelming evidence is that it was a contaminant resulting from a reduction in the carbon filtration and modification of the reverse osmosis process. The tragic incident does not reflect well on the testing procedures (or lack of them) by the firms that bought the bulk powder for tryptophan (or other pills) and converted it into "all natural" pills.

Conveniently ignored by the activists is their long history of opposing any regulation or even monitoring of these so-called dietary supplements and their ability to pass legislation, before and after the tragedy, which specifically exempts them from precisely the oversight that might have prevented it.In fact, in 1972 the U.S. Food and Drug Administration tried to withdraw the GRAS designation for amino acids in "dietary supplements because FDA felt that the available data did not allow" the FDA to deem them to be safe. The FDA had a number of concerns, including possible toxic or other adverse effects. Because of an error in its original GRAS designation, the FDA lost a court case that would have allowed it to have some reasonable oversight.

THE VITAMINS

Vitamins are natural, even if manufactured, as long as they are labeled organic. They too have not been tested for safety at the mega doses that some faddists use them, though clinical studies have shown that they can be helpful when used intelligently.

Obstetricians who frequently prescribe multiple vitamin pills for pregnant women and programmes for vitamin A intervention have saved children's eyesight and lives in poor countries. The faddists who would regulate agricultural chemicals out of existence oppose even simple truth-in-labeling requirements for their beloved natural products. Groups such as the International Advocates for Health Freedom, the Life Extension Foundation, and "a growing number of allies in the 'patriotic' movement" see a secret conspiracy of "international pharmaceutical empires working hand-in-hand with the UN and the FDA to restrict access to lowcost herbs, vitamins and minerals around the world".

ALFALFA SPROUTS

Alfalfa sprouts have been found to be contaminated with *E. coli* O157:H7 and salmonella. Obviously, not all alfalfa and other sprouts are contaminated but the rate of contamination for raw, untreated sprouts is considerably higher than for most other food.

The Emerging Infections Programme of the Centers for Disease Control has estimated that twenty thousand people in North America contracted salmonella from alfalfa spouts in 1995. Disease outbreaks have led to steps such as irradiating alfalfa seeds. Irradiation of alfalfa sprouts is currently being evaluated as an "adjunct seed decontamination method" Raw alfalfa and clover sprouts are recognized sources of foodborne illness in the United States.

Raw sprouts present unique food safety problems because the warm humid conditions needed to grow spouts are also ideal for the rapid growth of harmful bacteria. The very natural nature of alfalfa sprout production creates the contamination problem and makes them a "well-suited vehicle for salmonellosis." *JAMA* points out, "Seeds are often stored for months or years under cool, dry conditions in which salmonellae are stable.

During the three- to five-day sprouting process, numbers of salmonellae may increase by three to four orders of magnitude, decreasing little if at all during subsequent refrigeration." In addition, the way that alfalfa sprouts are stored, distributed, and marketed poses a danger to the health of the user: From farm to table, many opportunities exist for contamination of alfalfa seeds or sprouts. Crops can be easily contaminated with dirty water, run-off from adjacent farms, animal fertilizers used in previous growing seasons, or droppings from rodents or ruminants. The natural way that alfalfa sprouts are consumed at home or even in restaurants also adds to the danger.

"Alfalfa sprouts are rarely washed or cooked before consumption, and consumers are left with little protection other than chance". The salmonella organism is believed to be present in seed crevices between cotyledon and testa and therefore is not amenable to chemical treatment.At present there is a "fundamental problem" in that the commercial "sprouting process contains no 'kill step' that would eliminate pathogens without compromising a seed's germination potential. Consequently, because of the difficulties in producing a "pathogen-free seed," warnings have been issued about the dangers of eating alfalfa and other sprouts. The authors of the study conclude: "Alfalfa sprouts are a highrisk food for salmonellosis.

All consumers, particularly those at greatest risk for severe disease (immunosuppressed, elderly and very young people), should consider this danger when deciding whether to eat alfalfa sprouts". Proctor adds, "alfalfa sprouts contain 1.5 per cent by dry weight of canavine, a highly toxic substance". Not only is natural not necessarily better, food processing is an essential component of food safety. Says Garn, "Some cultivars are quite toxic, unless properly prepared".

QUANTIFYING THE IMPACTS ON FOOD SECURITY

Nanotechnologies enable food ingredients to be managed at a molecular level. They are used to create novel foods and nutriceuticals. Nanotechnology may also be used in food packaging to ensure, for example, better protection of food or to detect how fresh it is. However, to ensure consumer protection, specific properties and characteristics of nanomaterials need to be considered for any potential health risks, taking into account hazard assessment of nanotechnology products along the entire food chain (food, novel foods, food additives, food contact materials, feed, pesticides).

A number of studies have recently quantified the impacts of climate change on food security. In terms of quantifying agronomic yield change projections, these studies are either based on the AEZ tools developed by the IIASA analysis or the Decision Support System for Agrotechnology Transfer suite of crop models; all use the IIASA-Basic Linked System (BLS) economic model for assessing economic impacts. These tools, with some modifications relating to how crop yield changes are simulated, have also been used by others to undertake similar assessments and provide sensitivity analyses across a range of SRES and general circulation model (GCM) projections.

Many other simulations have also examined the effects of climate change with and without adaptation (induced technological progress, domestic policy change, international trade liberalization, etc.), with and without mitigation (*e.g.*, CO_2 stabilization, variants for temperature, rainfall change and distribution) or provide impact assessments for different speeds of climate change. The quantitative results for food security, trying to illuminate some of the differences and extract the main messages that emerge from the various studies. Unless indicated, all simulation results discussed below include the combined effects of climate change and elevated CO_2 on crops. The key messages can be summarized as follows:

First, it is very likely that climate change is likely to increase the number of people at risk of hunger compared with reference scenarios with no climate change; the exact impacts will, however, strongly depend on the projected socio-economic developments. For instance, it is estimated that climate change would increase the number of undernourished in 2080 by 5–26 per cent, compared with no climate change or by between 5 million and 10 million (B1 SRES) and 120 million to 170 million people (A2 SRES), with within-SRES ranges depending on GCM climate projections. Using only one GCM scenario, others projected small reductions by 2080, *i.e.*, –5 per cent, or –10 million (B1) to –30 million (A2) people, and slight increases of +13 per cent–26 per cent, or H–10 million (B2) to 30 million (A1) people.

Second, it is likely that the magnitude of these climate impacts will be small compared with the impact of socio-economic development. The limitations of socio-economic forecasts, these studies suggest that robust economic growth and a decline in population growth projected for the 21st century will, in all but one scenario (SRES A2), significantly reduce the number of people at risk of hunger in 2080. At any rate, the prevalence of undernourishment will decline as all scenarios assume that world population will continue to grow to 2080, albeit at lower rates.

Compared with FAO estimates of 820 million undernourished in developing countries today, several studies estimate reductions of >75 per cent by 2080, or H–560 million to 700 million people, projecting 100 million to 240 million undernourished by 2080 (A1, B1, and B2). As mentioned, the only exception is scenario A2, where the number of the hungry is forecast to decrease only slightly by 2080; but the higher population growth rates in A2 compared with other scenarios mean that the prevalence of undernourishment will decline drastically.

However, these analyses also confirm that the progress will be unevenly distributed over the developing world, and more importantly progress will be slow during the first decades of the outlook. With or without climate change, the millennium development goal of halving the prevalence of hunger by 2015 is unlikely to be realized before 2020–2030. In addition to socio-economic pressures considered within the IPCC SRES scenarios, food production may increasingly compete with bio-energy in coming decades; studies addressing possible consequences for world food supply have only started to surface, providing both positive and negative views. Importantly, none of the major world food models discussed herein have yet considered such competition.

Third, sub-Saharan Africa is likely to surpass Asia as the most food-insecure region. However, this is largely independent of climate change and is mostly the result of the socio-economic development paths assumed for the different developing regions in the SRES scenarios. Throughout most SRES and climate-change scenarios sub-Saharan Africa accounts for 40–50 per cent of global hunger by 2080, compared with H–24 per cent today; in some simulations sub-Saharan Africa even accounts for 70–75 per cent of global undernourishment by 2080.

These high estimates have emerged from slower growth variants of the A2 and B2 scenarios; also an A2 variant with slower population growth yields a sharper concentration of hunger in sub-Saharan Africa. For regions other than sub-Saharan Africa, results largely depend on GCM scenarios and therefore are highly uncertain.

Fourth, although there is significant uncertainty on the effects of elevated CO_2 on crop yields, this uncertainty is carried to a much lesser extent on food security. This result emerges from a comparison of climate change simulations with and without CO_2 fertilization effects on crop yields. As can be seen, higher CO_2 fertilization would not greatly affect global projections of hunger. In view of the fact that essentially all SRES worlds are characterized by much higher real incomes, much improved transportation and communication options, and still sufficient global food production, the somewhat smaller supplies will not be able to make a dent in global food security outcomes.

Many studies find that climate change without CO_2 fertilization would reduce the number of undernourished people by 2080 only by some 20 million to 140 million (120 million to 380 million for SRES A1, B1, and B2 without, compared with 100 million to 240 million with CO_2 fertilization effect). The exception again in these studies is SRES A2, under which the assumption of no CO_2 fertilization results in a projected range of 950 million to 1,300 million people undernourished in 2080, compared with 740 million to 850 million projected under climate change, but with CO_2 effects on crops.

Finally, recent research suggests large positive effects of climate stabilization for the agricultural sector. However, as the stabilizing effects of mitigation measures can take several decades to be realized from the moment of implementation, the benefits for crop production may be realized only in the second half of this century.

Importantly, even in the presence of robust global long-term benefits, regional and temporal patterns of winners and losers that can be projected with current tools are highly uncertain and depend critically on the underlying GCM projections.

UNCERTAINTIES AND LIMITATIONS IN FOOD SECURITY

The finding that socio-economic development paths have an important bearing on future food security and that they are likely to top the effects of climate change should not, or at least not only, be interpreted as a probability-based forecast. This is because SRES scenarios offer a range of possible outcomes "without any sense of likelihood". Yet SRES scenarios, like all scenarios, do not overcome the inability to accurately project future changes in economic activity, emissions, and climate.

Second, the existing global assessments of climate change and food security have only been able to focus on the impacts on food availability and access to food, without quantification of the likely important climate change effects on food safety and vulnerability (stability).This means that these assessments neither include potential problems arising from additional impacts due to extreme events such as drought and floods nor do they quantify the potential impacts of changes in the prevalence of food-borne diseases (positive as well as negative) or the interaction of nutrition and health effects through changes in the proliferation of vector-borne diseases such as malaria. On the food availability side, they also exclude the impacts of a possible sea-level rise for agricultural production or those that are associated with possible reductions of marine or freshwater fish production.

Third, it is important to note that even in terms of food availability, all current assessments of world food supply have focused only on the impacts of mean climate change, *i.e.*, they have not considered the possibility of significant shifts in the frequency of extreme events on regional production potential, nor have they considered scenarios of abrupt climate or socio-economic change; any of these scenario variants is likely to increase the already negative projected impacts of climate change on world food supply. Models that take into account the specific biophysical, technological, and market responses necessary to simulate realistic adaptation to such events are not yet available.

Fourth, this review finds that recent global assessments of climate change and food security rest essentially on a single modelling framework, the IIASA system, which combines the FAO/IIASA AEZ model with various GCM models and the IIASA BLS system, or on close variants of the IIASA system. This has important implications for uncertainty, given that the robustness of all these assessments strongly depends on the performance of the underlying models. There is, therefore, a clear need for continued and enhanced validation efforts of both the agro-climatological and food trade tools developed at IIASA and widely used in the literature. Finally, we note that assessments that do not only provide scenarios, but also attach probabilities for particular outcomes to come true could provide an important element for improved or at least better-informed policy decisions. A number of possibilities are offered to address the related modelling challenges. One option would be to produce such estimates with probability-based estimates of the (key) model parameters. Alternatively, the various scenarios could be constructed so that they reflect expert judgements on a particular issue. It would be desirable to attach probabilities to existing scenarios. Information on how likely the suggested outcomes are would contribute greatly to their usefulness for policy makers and help justify (or otherwise) policy measures to adapt to or mitigate the impacts of climate change on food security.

IMPACT OF CLIMATE CHANGE ON FOOD SECURITY

Climate change will affect all four dimensions of food security, namely food availability (*i.e.*, production and trade), access to food, stability of food supplies, and food utilization.

The importance of the various dimensions and the overall impact of climate change on food security will differ across regions and over time and, most importantly, will depend on the overall socio-economic status that a country has accomplished as the effects of climate change set in. Essentially all quantitative assessments show that climate change will adversely affect food security.

Climate change will increase the dependency of developing countries on imports and accentuate existing focus of food insecurity on sub-Saharan Africa and to a lesser extent on South Asia. Within the developing world, the adverse impacts of climate change will fall disproportionately on the poor. Many quantitative assessments also show that the socio-economic environment in which climate change is likely to evolve is more important than the impacts that can be expected from the biophysical changes of climate change. Less is known about the role of climate change for food stability and utilization, at least in quantitative terms.

However, it is likely that differences in socio-economic development paths will also be the crucial determinant for food utilization in the long run and that they will be decisive for the ability to cope with problems of food instability, be they climate-related or caused by other factors. Finally, all quantitative assessments we reviewed show that the first decades of the 21st century are expected to see low impacts of climate change, but also lower overall incomes and still a higher dependence on agriculture.

During these first decades, the biophysical changes as such will be less pronounced but climate change will affect those particularly adversely that are still more dependent on agriculture and have lower overall incomes to cope with the impacts of climate change. By contrast, the second half of the century is expected to bring more severe biophysical impacts but also a greater ability to cope with them. The underlying assumption is that the general transition in the income formation away from agriculture towards non-agriculture will be successful.

How strong the impacts of climate change will be felt over all decades will crucially depend on the future policy environment for the poor. Freer trade can help to improve access to international supplies; investments in transportation and communication infrastructure will help provide secure and timely local deliveries; irrigation, a promotion of sustainable agricultural practices, and continued technological progress can play a crucial role in providing steady local and international supplies under climate change.

9

Enviornmental Nanobiotechnology

The Chambers Science and Technology Dictionary defines biotechnology as 'the use of organisms or their components in industrial or commercial processes, which can be aided by the techniques of genetic manipulation in developing *e.g.* novel plants for agriculture or industry.' Despite the inclusiveness of this definition, the biotechnology sector is still often seen as largely medical or pharmaceutical in nature, particularly amongst the general public. While to some extent the huge research budgets of the drug companies and the widespread familiarity of their products makes this understandable, it does distort the full picture and somewhat unfairly so. However, while therapeutic instruments form, in many respects, the 'acceptable' face of biotechnology, elsewhere the science is all too frequently linked with unnatural interference.

While the agricultural, industrial and environmental applications of biotechnology are potentially very great, the shadow of Frankenstein has often been cast across them. Genetic engineering may be relatively commonplace in pharmaceutical thinking and yet in other spheres, like agriculture for example, society can so readily and thoroughly demonise it. The history of human achievement has always been episodic. For a while, one particular field of endeavour seems to hold sway as the preserve of genius and development, before the focus shifts and development forges ahead in dizzy exponential rush in an entirely new direction.

So it was with art in the renaissance, music in the 18th century, engineering in the 19th and physics in the 20th. Now it is the age of the biological, possibly best viewed almost as a rebirth, after the great heyday of the Victorian naturalists, who provided so much input into the developing science.

It is then, perhaps, no surprise that the European Federation of Biotechnology begins its 'Brief History' of the science in the year 1859, with the publication of On the Origin of Species by Means of Natural Selection by Charles Darwin. Though his famous voyage aboard HMS Beagle, which led directly to the formulation of his (then) revolutionary ideas, took place when he was a young man, he had delayed making them known until 1858, when he made a joint presentation before the Linnaean Society with Alfred Russell Wallace, who had, himself, independently come to very similar conclusions.

Their contribution was to view evolution as the driving force of life, with successive selective pressures over time endowing living beings with optimised characteristics for survival. Neo-Darwinian thought sees the interplay of mutation and natural selection as fundamental. The irony is that Darwin himself rejected mutation as too deleterious to be of value, seeing such organisms, in the language of the times, as 'sports' – oddities of no species benefit. Indeed, there is considerable evidence to suggest that he seems to have espoused a more Lamarckist view of biological progression, in which physical changes in an organism's lifetime were thought to shape future generations.

Darwin died in 1882. Ninety-nine years after his death, the first patent for a genetically modified organism was granted to Ananda Chakrabarty of the US General Electric, relating to a strain of Pseudomonas aeruginosa engineered to express the genes for certain enzymes in order to metabolise crude oil. Twenty years later still, in the year that saw the first working draft of the human genome sequence published and the announcement of the full genetic blueprint of the fruit fly, Drosophila melanogaster, that archetype of eukaryotic genetics research, biotechnology has become a major growth industry with increasing numbers of companies listed on the world's stock exchanges.

Thus, at the other end of the biotech timeline, a century and a half on from Origin of Species, the principles it first set out remain of direct relevance for what has been termed the 'chemical evolution' of biologically active substances and are commonly used in laboratories for in vitro production of desired qualities in biomolecules.

APPLICATIONS OF NANO-BIOTECHNOLOGY

Nanotechnology (prefix 'nano' means dwarf in Greek) is an amalgamation of science and engineering that has effected the measurment in the lowest units (nanometre - billionth of a metre). Nanotechnology is a valuable contribution of the profound resear ch by scientists and engineers to create new systems with novel functionalities. This gives nanotechnology greater share of control over a wide range of usage in applications right from high-density data to detecting DNA sequence to improved taste and wider variety of fruits and vegetables. On the commercial front, nanotechnology has enhanced every field in science, engineering, and technology; a phenomenal advantage, thus giving high efficiency and profitability.

Biotechnology is the application of biological research techniques to develop products that cater to the need of living beings.Nano-biotechnology, as the name suggests, is a mariage of nanotechnology and biotechnology. Each of these technologies has a tremendous impact by itself and therefore when they complement each other it results in efficacious fourth dimensional solutions.

Nano' is the present buzzword among science, technology and even business community. But is it a mere buzzword or has some real practical applications? For the answer to this question, we will have to travel in future. But, at the present times, all leading organizations have identified nanotechnology as their next frontier of research.

Nanotechnology is the technology working at extremely tiny scales (of nanometre dimensions). When you work at this level, you can manipulate atoms, influence chemical reactions and do many other unimaginable things with extremely high precision. It has been proposed that nanotechnology may lead to the production of nanobots which will be able to travel inside the body, interact with each other and even replicate. But, the design and engineering of these nanobots will be extremely tough and demanding. Does it mean that we will have to wait forever to see these nanobots in action?

The answer to the preceding question is simple: No! Nature has everything which we may desire in our so called nanobots. Living organisms can travel inside the body, generate their own power, communicate with each other and best of all, they can replicate. But, the one problem with living organisms which prevents their direct use as nanobots is that they are not able to interact with digital computers, which is a must for nanobots as they need to transfer information between human body and a computer. If nanotechnology can engineer silicon robots and biotechnology can provide replicating organisms, the natural way to realise both functions is to combine both of these fields. Voila! We have Nanobio-technology.

Nanobiotechnology is the merger of two distant fields of nanotechnology and biotechnology. It uses the concepts, components and processes from both of these fields to create new and novel applications. Already, basic research has began which is well demonstrated by the fact that ATP powered nanomotors have shown to be successfully working. Even though the commercial applications are still unfeasible, research into this field can yield surprises to what we believe is limits of technology.

The potential of Nanobiotechnology is immense. In future, it may be used to engineer nanobots which can act as personalized doctors acting round the clock. Or it may be used to create tiny cell-bombs which seek and destroy tumor cells while making sure that healthy cells are unaffected. Apart from medical applications, this technology may be employed to harvest energy directly from photosynthetic bacteria and convert it into electricity.

But before Nanobiotechnology realises its full potential, there are many hurdles to overcome. First and foremost is offered by our body itself. The vertebrate immune system battles most of the foreign particles called antigens (such as pathogens, etc.) by producing antibodies which render them inactive. Our body's immune system will also classify nano-machines inside the body as foreign particles (which they actually are), hence these nano-machines may not work in our bodies unless some special modifications are made.

Ethical and moral issues also haunt this nascent field. Is it ethical to put foreign particles inside the human body? The horror of self-replicating nano-machines going wild is also a debatable issue. This added to fact that since these nano-machines will have biological components, they will also be undergoing natural selection from evolution, which can result in new entities capable of replicating and harming like viruses, bacteria or in their own novel ways.

Once all these issues get fixed up, expect your life to be radically changed. Nanobiotechnology may bring immense paradigm shift that we may wonder that how did we live without it! Expect Nanobiotechnology to be for 21st century what Electronics was for 20th century and what Renaissance was for 15th century.

Nano technology is a relatively new technology which came into its own around the year 2000, Nano technology is a step into the future of miniaturization. In today's world smaller is better, examples of this are seen with electronics equipment. And the best example of this? A few years ago home computers were big and bulky, so were monitors.

Today we have slim line wafer thin monitors with flat screens and much slimmer computer bases with more power than ever before, due to the smaller circuits boards inside. The word Nano technology derives from the word Nano metre which is one thousandth of a micron, which is roughly the size of a single molecule.

Nano technology is the study, manufacture and manipulation of any small structure or machine that can be made of as little as one molecule. Nano technology has only been made possible by the advancement in microscopic tools designed for looking at and allowing manipulation of single molecules.

For those studying molecular biology, Nano biotechnology is a major breakthrough, the molecular biologist joins forces with the Nano technologist to get a better understanding and gain access to the nanostructures and Nano machines which were designed by over 4 billion years of engineering, cell machinery and biological molecules.

Today Nano technologists are set to accomplish many different goals and achieve many things that wouldn't otherwise be possible by exploiting biological molecules and cell processes. At one time silicon scaffolding would be used as the basis for Nano structures however DNA's ladder structure provides Nano biotechnologists with the perfect natural framework from which to assemble Nano structures, with its highly specific bonding properties it is the perfect base for bringing together atoms in a predictable pattern to create a Nano structure.

Nano technologists have also come to rely on the biological molecules self-assembling properties when it comes to the creation of Nano structures, such as the lipids which are able to form liquid crystals spontaneously.

DNA is not only being put to good use in building Nano structures but is also a very essential component when it comes to building Nano machines, DNA which is an information storage molecule could very well in the future serve as the basis for the next generation of computers.

As mentioned earlier microprocessors and microcircuits have already shrunk over the last few years allowing smaller machinery, it is quite possible they will in the future shrink to Nano processors and Nano microcircuits.

This is not as far fetched as it might sound as DNA molecules can be mounted onto microchips and may eventually come to replace the microchips with electron flow channels etched in silicon. Biochips such as theses are DNA based processors that can take advantage of and use DNA's extraordinary capacity to store information.

Biochips are able to exploit the properties of DNA in order to solve complex computational problems, this put very simple means that they are used to do math. Scientific studies have show us that 1,000 DNA molecules were able to solve complex computational problems that would have taken over a century for a computer to solve, in just four months.

Other biological molecules are helping in many different ways as biologists strive towards being able to store and transmit more and more information in smaller places. A good example of this is the common CD that hold information and music, researchers are now using light absorbing molecules very similar to that found in our retinas to increase the storage capacity of a single CD by, a thousand fold.

NANOBIOTECHNOLOGY AT ITS BEST

Technology is turning around to help the humans by providing easier methods to keep up a healthy living. As the population in the modern times ages, scientists are finding it difficult to manage the right combination of drugs to make them effective. The existing means of intake of drugs orally or through injections are becoming moderately effective and can be extremely painful due to the non-availability of accurate diagnosis.

One of the newer methods to deliver drugs at the right place, time, and amounts is being addressed through nano-biotechnology. Scientists and engineers have come out with new systems where the human-body parametres are monitored using a semiconductor chip that is placed inside a human body.

The information recorded/acquired is then transmitted to another wrist-watch like equipment using telemetry, which, in turn, instructs the semiconductor chip to deliver the right amounts of drug to the right place.

For example: When a patient is being treated for diabetes, the semiconductor chip monitors the blood-sugar levels and passes on the information to the device outside of the body. The external device, in turn, instructs the semiconductor chip that acts as the drug delivery system, to release the right amounts of insulin into the blood, thus causing less or no side effects to the patient.

Nanotechnology used in biotechnology is definitely a life-saving innovation that has come as a boon to mankind. The above semiconductor chip application is just a modicum of what it is capable in terms of its impact on human race. The real applications can turn around the life-cycle of human beings in the years to come.

The method we are currently studying is dielectr-ophoresis. This physical effect allows voltages on microfabricated electrodes (which we fabricate in the Integrated Nanosystems Research Facility) to manipulate living cells in a solution through electric fields, without ever coming into contact with the cells themselves.

This effect may be useful for on-chip preparation of biological and chemical specimens before further chemical, electric, and optical analysis occurs, presumably on the same chip. Although many years away, the integration of all these functions onto a single chip will ultimately allow cost effective mass production of a new generation of diagnostic and analytical tools.

In figure, the principle is illustrated schematically. If a polarizable object is placed in an electric field, there will be an induced positive charge on one side of the object and induced negative charge (of the same magnitude as the induced positive charge) on the other side of the object. The positive charge will experience a "pulling" force; the negative charge will experience a "pushing" force. In a uniform field, as depicted in figure a, the "pulling" force will cancel the "pushing" force, and the net force will be zero.

However, in a non-uniform field, as depicted in figure b, the electric field will be stronger on one side of the object and weaker on the other side of the object. Hence, the "pulling" and "pushing" forces will not cancel, and there will be a net force on the object.

Uniform field

Net force = 0

Non-uniform field

Net force $\neq 0$

A

B

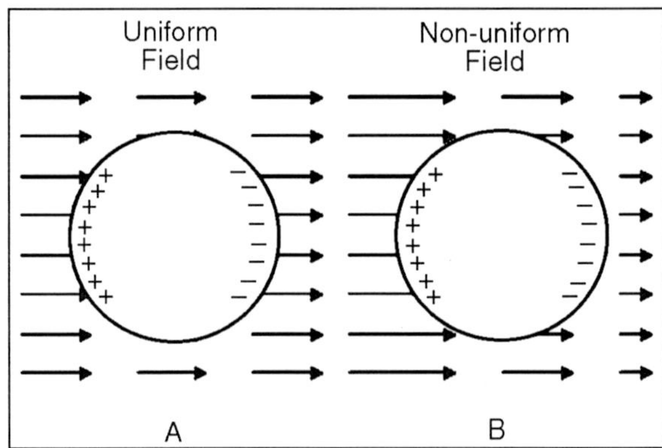

Fig. Schematic Depiction of Di-electrophoresis.

The direction of the arrows represents the direction of the electric field; the length of the arrows represents the magnitude of the electric field. In reality, the electric field lines would bend around the sphere somewhat. We neglect that effect for this simple depiction.

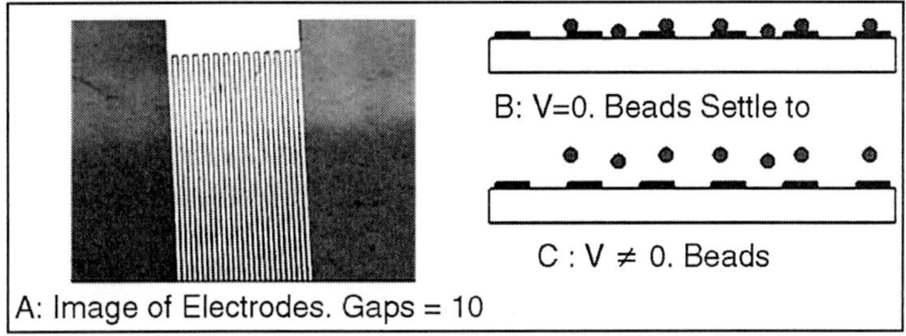

Fig. First Generation of UCI Dielectrophoresis Experiments.

A: Image of electrodes. Gaps = 10m.
B: V = 0. Beads settle to bottom.
C: V ≠ 0. Beads"levitate".

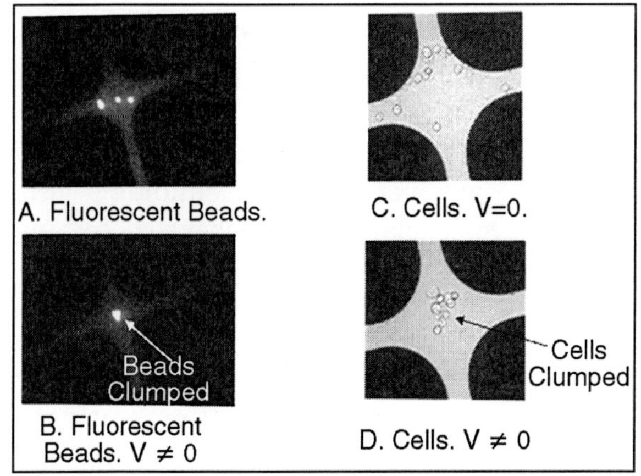

Fig. Second Generation of UCI Dielectrophoresis Experiments. There are four Electrodes which Surround the trap. The central Region is Approximately 100 m across.

A: Fluorescent beads. V= 0.
B: Fluorescent beads. V ≠ 0.
C: Cells. V=0.
D: Cells. V ≠ 0.
Beads clumped in"trap"
Cells clumped in"trap"

Nanobiotechnology is the application of nanotechnology for the study of biological and biomedical systems. One of the major areas is the utility of nanoscale systems and nanofabricated devices to guide stem cell development, mimic tissue regeneration and develop tools for cancer research.

Since stem cells are able to self-renew and differentiate into specialized cell types, so called pluripotency, they have great potential to better understand diseases, aid in drug discovery and treat diseases. The breakthrough findings that adult cells can be reprogrammed to Induced Pluripotent Stem Cells (iPSCs) provide an important cell source for regenerative medicine and drug discovery. Using a set of reprogramming factors linked to pluripotency, iPSCs can be produced from normal human cells or patient's cells with specific diseases. These iPSCs hold great promise for the study and treatment of human disease. Nanofabrication-based gene delivery of reprogramming factors into cells will facilitate generating iPSCs in a safe, efficient and effective manner.

The self-renewal and differentiation of stem cells are controlled by stem cell microenvironment which constitutes of extracellular matrices, soluble factors and neighbouring cells. The extracellular matrix is an interconnected fibrous network with nanoscale architecture. Most of the soluble factors are growth factors which are naturally occurring proteins. Cell-cell junctions play a critical role in stem cell proliferation and differentiation. It is essential to understand how stem cells interact with their microenvironment before realizing the full potential of iPSCs and other stem cells. The nanoscale interactions between stem cells and their microenvironment can be revealed by metrology tools such as scanning electron microscopy, transmission electron microscopy, atomic force microscopy and fluorescence microscopy combined with cell biology techniques. The advance in nanotechnology will provide better metrology tools as well as molecular probes to visualize cell-microenvironment interactions.

Cells and tissues in the body are three-dimensional (3-D) architectures with nanoscale interactions. Stem cells are usually studied in vitro using two-dimensional (2-D) monolayer culture. In many cases, cells grown in 2-D culture fail to recapitulate some of the key stem cell properties. Therefore, it is necessary to establish 3-D tissue models for stem cell studies. Using the tissue engineering principle, 3-D cultivation systems can be established to mimic stem cell microenvironments, propagate stem cells and differentiate them into cell types of interest. The stem cell microenvironment can be manipulated with nanofibrous matrices or surfaces with nanotopography to mimic the extracellular matrix and basement membrane, nanoparticles or micro-devices to control the release of soluble factors, cell deposition techniques to place neighbouring cells close to stem cells. The ability to reconstruct stem cell microenvironments will facilitate the understanding of stem cell-microenvironment interactions and the maintenance of stem cell properties in vitro. Micro-/Nanofabrication technologies make it possible to construct micro-tissue models for understanding stem cell differentiation and tissue morphogenesis. The drugs that regulate stem cell behaviours can be good therapeutics. In this way, the 3-D micro-tissue model provides a platform for screening therapeutics and testing drug efficacy.

A rare subtype of cancer cells shares similarities with normal stem cells which possess unlimited self-renewal potential and differentiation capacity. These cells are referred to cancer stem cells. Comparing behaviours of embryonic stem cells, iPSCs generated from patients with cancer, and cancer stem cells in 3-D cultivation system will facilitate the discovery of therapeutic target. The surrounding microenvironment of cancer cells plays a critical role in tumorigenesis and metastasis. In addition to extracellular matrix (ECM) and soluble factors, cancer cell microenvironment is a complex system composed of many cell types. Stem cells can be used to mimic the cellular components of cancer cell microenvironment. Micro-/nanotechnology can be used to construct 3-D microenvironments and deliver diagnostic and therapeutic agents. The established 3-D tumor model will bridge the conventional 2-D culture and animal experiments for cancer research, anti-cancer drug screening, and novel diagnostic and therapeutics testing.

In summary, the marriage of nanotechnology and stem cell biology will provide a new avenue to address challenges in cancer research. This exciting frontier is expected to lead to transformative change in the way we diagnose and treat cancer.

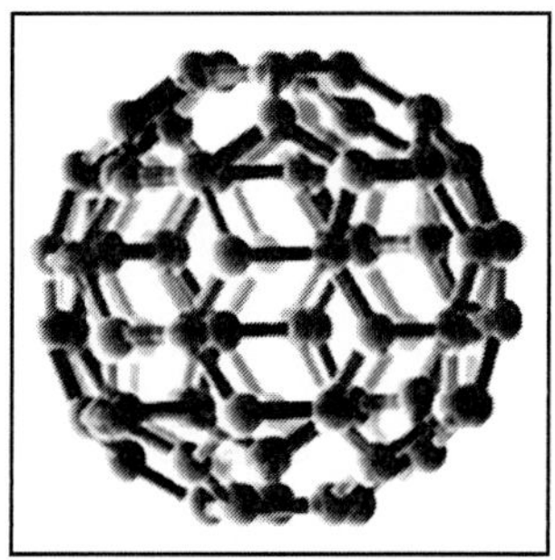

Nanobiotechnology is a new science involved in the design of tools and processes for use in the study of biosystems. These tools and processes are derived from another fairly new field called nanotechnology. Nanobiotechnology has the potential to create a new generation of medical research and practice.

THE ROLE OF ENVIRONMENTAL BIOTECHNOLOGY

While pharmaceutical biotechnology represents the glamorous end of the market, environmental applications are decidedly more in the Cinderella mould. The reasons for this are fairly obvious. The prospect of a cure for the many diseases and conditions currently promised by gene therapy and other biotech-oriented medical miracles can potentially touch us all. Our lives may, quite literally, be changed.

Environmental biotechnology, by contrast, deals with far less apparently dramatic topics and, though their importance, albeit different, may be every bit as great, their direct relevance is far less readily appreciated by the bulk of the population. Cleaning up contamination and dealing rationally with wastes is, of course, in everybody's best interests, but for most people, this is simply addressing a problem which they would rather had not existed in the first place.

Even for industry, though the benefits may be noticeable on the balance sheet, the likes of effluent treatment or pollution control are more of an inevitable obligation than a primary goal in themselves. In general, such activities are typically funded on a distinctly limited budget and have traditionally been viewed as a necessary inconvenience. This is in no way intended to be disparaging to industry; it simply represents commercial reality. In many respects, there is a logical fit between this thinking and the aims of environmental biotechnology. For all the media circus surrounding the grand questions of our age, it is easy to forget that not all forms of biotechnology involve xenotransplantation, genetic modifica-tion, the use of stem cells or cloning.

Some of the potentially most beneficial uses of biological engineering and which may touch the lives of the majority of people, however indirectly, involve much simpler approaches. Less radical and showy, certainly, but powerful tools, just the same. Environmental biotechnology is fundamentally rooted in waste, in its various guises, typically being concerned with the remediation of contamination caused by previous use, the impact reduction of current activity or the control of pollution.

Thus, the principal aims of this field are the manufacture of products in environmentally harmonious ways, which allow for the minimisation of harmful solids, liquids or gaseous outputs or the clean-up of the residual effects of earlier human occupation. The means by which this may be achieved are essentially two-fold.

Environmental biotechnologists may enhance or optimise conditions for existing biological systems to make their activities happen faster or more efficiently, or they resort to some form of alteration to bring about the desired outcome. The variety of organisms which may play a part in environmental applications of biotechnology is huge, ranging from microbes through to trees and all are utilised on one of the same three fundamental bases – accept, acclimatise or alter. For the vast majority of cases, it is the former approach, accepting and making use of existing species in their natural, unmodified form, which predominates.

The Scope for Use

There are three key points for environmental biotechno-logy interventions, namely in the manufacturing process, waste management or pollution control. Accordingly, the range of businesses to which environmental biotechnology has potential relevance is almost limitless. One area where this is most apparent is with regard to waste.

All commercial operations generate waste of one form or another and for many, a proportion of what is produced is biodegradable. With disposal costs rising steadily across the world, dealing with refuse constitutes an increasingly high contribution to overheads. Thus, there is a clear incentive for all businesses to identify potentially cost-cutting approaches to waste and employ them where possible.

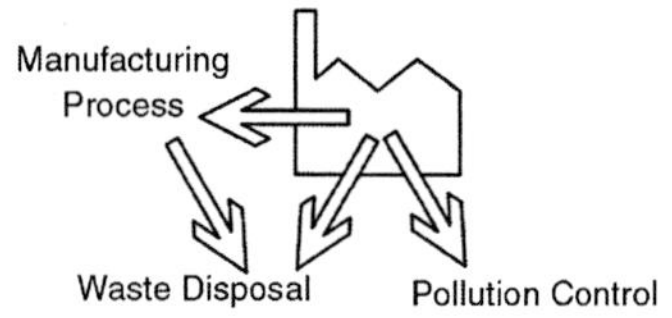

Fig. The Three Intervention Points

Changes in legislation throughout Europe, the US and elsewhere, have combined to drive these issues higher up the political agenda and biological methods of waste treatment have gained far greater acceptance as a result. For those industries with particularly high biowaste production, the various available treatment biotechnologies can offer considerable savings. Manufacturing industries can benefit from the applications of whole organisms or isolated biocomponents. Compared with conventional chemical processes, microbes and enzymes typically function at lower temperatures and pressures.

The lower energy demands this makes leads to reduced costs, but also has clear benefits in terms of both the environment and workplace safety. Additionally, biotechno-logy can be of further commercial significance by converting low-cost organic feedstocks into high value products or, since enzymatic reactions are more highly specific than their chemical counterparts, by deriving final substances of high relative purity.

Almost inevitably, manufacturing companies produce wastewaters or effluents, many of which contain biodegrada-ble contaminants, in varying degrees. Though traditional permitted discharges to sewer or watercourses may be adequate for some, other industries, particularly those with recalcitrant or highly concentrated effluents, have found significant benefits to be gained from using biological treatment methods themselves on site. Though careful monitoring and process control are essential, biotechnology stands as a particularly cost-effective means of reducing the pollution potential of wastewater, leading to enhanced public relations, compliance with environmental legislation and quantifiable cost-savings to the business.

Those involved in processing organic matter, for example, or with drying, printing, painting or coating processes, may give rise to the release of volatile organic compounds (VOCs) or odours, both of which represent environmental nuisances, though the former is more damaging than the latter. For many, it is not possible to avoid producing these emissions altogether, which leaves treating them to remove the offending contaminants the only practical solution. Especially for relatively low concentrations of readily water-soluble VOCs or odorous chemicals, biological technologies can offer an economic and effective alternative to conventional methods.

The use of biological cleaning agents is another area of potential benefit, especially where there is a need to remove oils and fats from process equipment, work surfaces or drains. Aside from typically reducing energy costs, this may also obviate the need for toxic or dangerous chemical agents. The pharmaceutical and brewing industries, for example, both have a long history of employing enzyme-based cleaners to remove organic residues from their process equipment. In addition, the development of effective biosensors, powerful tools which rely on biochemical reactions to detect specific substances, has brought benefits to a wide range of sectors, including the manufacturing, engineering, chemical, water, food and beverage industries. With their ability to detect even small amounts of their particular target chemicals, quickly, easily and accurately, they have been enthusiastically adopted for a variety of process monitoring applications, particularly in respect of pollution assessment and control.

Contaminated land is a growing concern for the construction industry, as it seeks to balance the need for more houses and offices with wider social and environmental goals. The reuse of former industrial sites, many of which occupy prime locations, may typically have associated planning conditions attached which demand that the land be cleaned up as part of the development process.

With urban regeneration and the reclamation of 'brown-field' sites increasingly favoured in many countries over the use of virgin land, remediation has come to play a significant role and the industry has an ongoing interest in identifying cost-effective methods of achieving it. Historically, much of this has involved simply digging up the contaminated soil and removing it to landfill elsewhere.

Bioremediation technologies provide a competitive and sustainable alternative and in many cases, the lower disturbance allows the overall scheme to make faster progress. As the previous brief examples show, the range of those which may benefit from the application of biotechnology is lengthy and includes the chemical, pharmaceutical, water, waste management and leisure industries, as well as manufacturing, the military, energy generation, agriculture and horticulture.

Clearly, then, this may have relevance to the viability of these ventures and, as was mentioned at the outset, biotechno-logy is an essentially commercial activity. Environmental biotechnology must compete in a world governed by the best practicable environmental option (BPEO) and the best available techniques not entailing excessive cost (Batneec).

Consequently, the economic aspect will always have a large influence on the uptake of all initiatives in environmental biotechnology and, most particularly, in the selection of methods to be used in any given situation. It is impossible to divorce this context from the decisionmaking process. By the same token, the sector itself has its own implications for the wider economy.

ENVIRONMENT POLICY AND ENVIRONMENT MANAGEMENT SYSTEM

Driven by its commitment for sustainable growth of power, NTPC has evolved a well defined environment management policy and sound environment practices for minimising environmental impact arising out of setting up of power plants and preserving the natural ecology.

NATIONAL ENVIRONMENT POLICY

At the national level, the Ministry of Environment and Forests had prepared a draft Environment Policy (NEP) and the Ministry of Power along with NTPC actively participated in the deliberations of the draft NEP. The NEP 2006 has since been approved by the Union Cabinet in May 2006.

NTPC ENVIRONMENT POLICY

As early as in November 1995, NTPC brought out a comprehensive document entitled "NTPC Environment Policy and Environment Management System". Amongst the guiding principles adopted in the document are company's proactive approach to environment, optimum utilisation of equipment, adoption of latest technologies and continual environment improvement. The policy also envisages efficient utilisation of resources, thereby minimising waste, maximising ash utilisation and providing green belt all around the plant for maintaining ecological balance.

ENVIRONMENT MANAGEMENT, OCCUPATIONAL HEALTH AND SAFETY SYSTEMS

NTPC has actively gone for adoption of best international practices on environment, occupational health and safety areas. The organization has pursued the Environmental Management System (EMS) ISO 14001 and the Occupational Health and Safety Assessment System OHSAS 18001 at its different establishments. As a result of pursuing these practices, all NTPC power stations have been certified for ISO 14001 and OHSAS 18001 by reputed national and international Certifying Agencies.

POLLUTION CONTROL SYSTEMS

While deciding the appropriate technology for its projects, NTPC integrates many environmental provisions into the plant design. In order to ensure that NTPC comply with all the stipulated environment norms, various state-of-the-art pollution control systems/ devices as discussed below have been installed to control air and water pollution.

Electrostatic Precipitators

The ash left behind after combustion of coal is arrested in high efficiency Electrostatic Precipitators (ESPs) and particulate emission is controlled well within the stipulated norms. The ash collected in the ESPs is disposed to Ash Ponds in slurry form.

Flue Gas Stacks

Tall Flue Gas Stacks have been provided for wide dispersion of the gaseous emissions (SOX, NOX etc) into the atmosphere.

Low-NOX Burners

In gas based NTPC power stations, NOx emissions are controlled by provision of Low-NOx Burners (dry or wet type) and in coal fired stations, by adopting best combustion practices.

Neutralisation Pits

Neutralisation pits have been provided in the Water Treatment Plant (WTP) for pH correction of the effluents before discharge into Effluent Treatment Plant (ETP) for further treatment and use.

Coal Settling Pits/ Oil Settling Pits

In these Pits, coal dust and oil are removed from the effluents emanating from the Coal Handling Plant (CHP), coal yard and Fuel Oil Handling areas before discharge into ETP.

DE and DS Systems

Dust Extraction (DE) and Dust Suppression (DS) systems have been installed in all coal fired power stations in NTPC to contain and extract the fugitive dust released in the Coal Handling Plant (CHP).

Cooling Towers

Cooling Towers have been provided for cooling the hot Condenser cooling water in closed cycle Condenser Cooling Water (CCW) Systems. This helps in reduction in thermal pollution and conservation of fresh water.

ASH DYKES AND ASH DISPOSAL SYSTEMS

Ash ponds have been provided at all coal based stations except Dadri where Dry Ash Disposal System has been provided. Ash Ponds have been divided into lagoons and provided with garlanding arrangements for change over of the ash slurry feed points for even filling of the pond and for effective settlement of the ash particles. Ash in slurry form is discharged into the lagoons where ash particles get settled from the slurry and clear effluent water is discharged from the ash pond. The discharged effluents conform to standards specified by CPCB and the same is regularly monitored. At its Dadri Power Station, NTPC has set up a unique system for dry ash collection and disposal facility with Ash Mound formation. This has been envisaged for the first time in Asia which has resulted in progressive development of green belt besides far less requirement of land and less water requirement as compared to the wet ash disposal system.

ASH WATER RECYCLING SYSTEM

Further, in a number of NTPC stations, as a proactive measure, Ash Water Recycling System (AWRS) has been provided. In the AWRS, the effluent from ash pond is circulated back to the station for further ash sluicing to the ash pond. This helps in savings of fresh water requirements for transportation of ash from the plant.

The ash water recycling system has already been installed and is in operation at Ramagundam, Simhadri, Rihand, Talcher Kaniha, Talcher Thermal, Kahalgaon, Korba and Vindhyachal. The scheme has helped stations to save huge quantity of fresh water required as make-up water for disposal of ash.

Dry Ash Extraction System (DAES)

Dry ash has much higher utilization potential in ash based products (such as bricks, aerated autoclaved concrete blocks, concrete, Portland pozzolana cement, etc.). DAES has been installed at Unchahar, Dadri, Simhadri, Ramagundam, Singrauli, Kahalgaon, Farakka, Talcher Thermal, Korba, Vindhyachal, Talcher Kaniha and BTPS.

Liquid Waste Treatment Plants and Management System

The objective of industrial liquid effluent treatment plant (ETP) is to discharge lesser and cleaner effluent from the power plants to meet environmental regulations. After primary treatment at the source of their generation, the effluents are sent to the ETP for further treatment. The composite liquid effluent treatment plant has been designed to treat all liquid effluents which originate within the power station *e.g.* Water Treatment Plant (WTP), Condensate Polishing Unit (CPU) effluent, Coal Handling Plant (CHP) effluent, floor washings, service water drains etc.

The scheme involves collection of various effluents and their appropriate treatment centrally and re-circulation of the treated effluent for various plant uses.

NTPC has implemented such systems in a number of its power stations such as Ramagundam, Simhadri, Kayamkulam, Singrauli, Rihand, Vindhyachal, Korba, Jhanor Gandhar, Faridabad, Farakka, Kahalgaon and Talcher Kaniha. These plants have helped to control quality and quantity of the effluents discharged from the stations.

Sewage Treatment Plants and Facilities

Sewage Treatment Plants (STPs) sewage treatment facilities have been provided at all NTPC stations to take care of Sewage Effluent from Plant and township areas. In a number of NTPC projects modern type STPs with Clarifloculators, Mechanical Agitators, sludge drying beds, Gas Collection Chambers etc have been provided to improve the effluent quality.

The effluent quality is monitored regularly and treated effluent conforming to the prescribed limit is discharged from the station. At several stations, treated effluents of STPs are being used for horticulture purpose.

ENVIRONMENTAL INSTITUTIONAL SET-UP

Realizing the importance of protection of the environment with speedy development of the power sector, the company has constituted different groups at project, regional and Corporate Centre level to carry out specific environment related functions.

The Environment Management Group, Ash Utilisation Group and Centre for Power Efficiency and Environment Protection (CENPEEP) function from the Corporate Centre and initiate measures to mitigate the impact of power project implementation on the environment and preserve ecology in the vicinity of the projects. Environment Management and Ash Utilisation Groups established at each station, look after various environmental issues of the individual station.

Environment Management During Operation Phase

NTPC's environment friendly approach to power has already begun to show results in conservation of natural resources such as water and fuel (coal, oil and gas) as well as control of environmental pollution. As already mentioned earlier, NTPC has chalked out a set of well defined activities that are envisaged right from the project conceptualisation stage so that during the entire life cycle of the power plant, NTPC is fully compliant with various environment regulations and a pristine environment and ecological balance is maintained in and around its power station and townships.

Following is brief description of some of the measures taken during the operation phase of the stations. Performance enhancement and up-gradation measures are undertaken by the organisation during the post operational stage of the stations. These activities have greatly helped to minimise the impact on environment and preserve the ecology in and around its power projects. These measures have been enumerated as follows.

Monitoring of Environmental Parameters

A broad based Environment Monitoring Programme has been formulated and implemented in NTPC. All pollutants discharged from the power plant such as stack emission, ash pond effluent, main plant effluent, domestic effluent and Condenser Cooling Water (CCW) effluent are monitored at the stipulated frequency at the source itself and at the points of discharge. In addition to the above, ambient air, surface water and ground water quality in and around NTPC plants are regularly monitored to assess any adverse impacts as a result of operation of the power plant.

On-Line Data Base Management

In order to have better control on pollution and to achieve effective environment management in and around NTPC stations, it is imperative to have an on-line, reliable and efficient environment information system on the plant operational and environmental performance parameters at all three levels i.e generating Stations, Regional Headquarters and Corporate Centre. In consideration of the above, a computerized programme, namely "Paryavaran Monitoring System" - PMS, which could provide reliable storage, prompt and accurate flow of information on environmental performance of Stations was developed and installed in NTPC.

This software facilitates direct transfer of environment reports and other environment related information from stations to the Regional Headquarters and Corporate Centre. The PMS has already been implemented at Corporate Centre, the Regional Headquarters and most of the Stations. This system has helped in achieving continuous improvement in NTPCs environment performance through improved monitoring and reporting system by using the trend analysis and advanced data management techniques.

ENVIRONMENT REVIEWS

To maintain constant vigil on environmental compliance, Environmental Reviews are carried out at all operating stations and remedial measures have been taken wherever necessary. As a feedback and follow-up of these Environmental Reviews, a number of retrofit and up-gradation measures have been undertaken at different stations. Such periodic Environmental Reviews and extensive monitoring of the facilities carried out at all stations have helped in compliance with the environmental norms and timely renewal of the Air and Water Consents.

Upgradation and Retrofitting of Pollution Control Systems

In order to keep pace with the changing norms and ensure compliance with statutory requirements in the field of pollution control, NTPC keeps an open mind for Renovation and Modernisation (R & M) and Retrofitting and Upgradation of pollution monitoring and control facilities in its existing stations. It is important to mention that such modifications/retrofit programmes not only helped in betterment of environment but also in resource conservation. High efficiency Electro-Static Precipitators (ESPs) of the order of 99.5 per cent and above have been provided at NTPC stations for control of stack particulate emissions.

However, the ESPs of a number of stations were built prior to the promulgation of the Environment (Protection) Act, 1986 and notification of emission control standards under this Act. Remedial measures have already been taken up and implemented to improve the efficiency of the existing ESPs at various NTPC stations. ESP performance enhancement programme by adopting advanced microprocessor based Electrostatic Precipitator Management System (EPMS) was installed at its power stations at Singrauli, Ramagundam,

Korba, Farakka, Rihand, Vindhyachal and Unchahar. Additional ESPs were retrofitted in the older power stations, namely at Badarpur and Talcher Thermal.

As a result of the above retrofits, the emission of Suspended Particulate Matter (SPM) has been brought down appreciably at the above stations and is maintained within the present statutory limit of 150 mg/Nm^3. In new projects, the ESPs have been designed for a maximum permissible outlet dust emission of 50 mg/Nm^3 to meet the likely stringent emission norms in the near future.

RESOURCES CONSERVATION

With better awareness and appreciation towards ecology and environment, the organization is continually looking for innovative and cost effective solutions to conserve natural resources and reduce wastes.

Some of the measures include:

- Reduction in land requirements for main plant and ash disposal areas in newer units.
- Capacity addition in old plants, within existing land.
- Reduction in water requirement for main plant and ash disposal areas through recycle and reuse of water.
- Efficient use of Fuel (Coal, Natural gas and Fuel oil)
- Reduction in fuel requirement through more efficient combustion and adoption of state-of-the-art technologies such as super critical boilers

WASTE MANAGEMENT

Various types of wastes such as Municilal or domestic wastes, hazardous wastes, Bio-Medical wastes get generated in power plant areas, plant hospital and the townships of projects. The wastes generated are a number of solid and hazardous wastes like used oils and waste oils, grease, lead acid batteries, other lead bearing wastes (such as garkets etc.), oil and clarifier sludge, used resin, used photochemicals, asbestos packing, e-waste, metal scrap, C&I wastes, electricial scrap, empty cylinders (refillable), paper, rubber products, canteen (bio-degradable) wastes, buidling material wastes, silica gel, glass wool, fused lamps and tubes, fire resistant fluids etc.

These wastes fall either under hazardous wastes category or non-hazardous wastes category as per classification given in Government of Indias notification on Hazardous Wastes (Management and Handling) Rules 1989 (as amended on 06.01.2000 and 20.05.2003). Handling and manegement of these wastes in NTPC stations have been discussed below.

Municipal Waste Management

Domestic or municipal waste is generated in households at townships. This waste is seggregated into bio-degradable and nonbiodegradable wastes at source itself in different coloured containers and thereafter the two types are disposed separately. Bio-degradable waste is spread uniformly in identified low lying areas and thereafter it is covered with soil for use later as manure after composting. The seggregated non- bio-degradable waste is disposed off separately in other identified low lying areas and is spread out uniformly.

Hazardous Waste Management

NTPC being a proactive organization, the handling and disposal of hazardous wastes are done as per the Hazardous Wastes (Management and Handling) Rules 1989 (as amended in 2003) guidelines issued by Government of India for the treatment, storage and disposal of hazardous wastes. Scientific study on management and handling of hazardous wastes was carried out at a few NTPC stations to adopt the best practices so that there is a complete compliance with statutory requirements. In NTPC sataione, the Hazardous Wastes (Recyclable) are sold/ auctioned to registered recyclers/ refiners.

The other hazardous wastes such as the activated carbon resins, used drums (hazardous) chromium (Cr-III electrolytes, used petro-chemicals, asbestos packings, used torch batteries, ribbon, toners/ cartridges, mixed wastes (waste oil, water and cotton) filters, earth contaminated with synthetic oil (FQF) glass used and sodium silicate, lamps and tubes etc. fall under the category of Hazardous Wastes (Non-Recyalable).

These wastes are small in quantity and are stored in properly identified locations. As per the notification, hazardous wastes (non-recyalable) are to be sent to State Pollution Control Board (SPCB) approved common treatment storage and disposal facility (TSDF).

Bio-Medical Waste Management

Hospital (or Bio-medical) wastes get generated from hospitals and they include urine bags, human anatomical wastes, plaster of aris waste, empty plastic bottles of water and glucose, blood and chemical mixed cotton, blood and urines tubes etc. these wastes are segregated and are placed in buckets of different colours as per the notification for Bio-Medical Waste (Management and Handling) Rules.

The seggregated bio-medical wastes are either disposed through the SPCB approved agency or they are treated in autoclaves before disposal into bio-medical waste disposal pits. The treated bio-medical waste is spread uniformly and covered with 10 cm thick soil in bio-medical waste disposal pits.

LAND USE/ BIO-DIVERSITY

As a policy, NTPC lays special emphasis on land use and Bio-diversity by way of development of green belts, energy plantations, reclmation of abandoned Ash Ponds and EIA and ecological monitoring in the project areas and its surroundings.

Reclamation of Abandoned Ash ponds

The reclamation of abandoned ash pond sites is a challenging task. NTPC has reclaimed temporary ash disposal areas at some of its projects namely Ramagundam, Talcher Thermal, Rihand, Singrauli and Unchahar through plantation and converted these sites into lush green environments. Extensive plantations have also been undertaken on dry ash mound at NTPC-Dadri. It is planned to reclaim all the abandoned ash disposal areas by plantation.

Green Belts, Afforestation and Energy Plantations

Whats more, in a concerted bid to counter the growing ecological threat, NTPC is undertaking afforestation programmes covering vast areas of land in and around its projects. Appropriate afforestation programmes for plant, township and green belt areas of the project have been implemented at all projects. In order to enhance green cover in the areas around our projects, as a responsible corporate citizen, NTPC till date has planted more than 18 million trees at its projects throughout the country.

The afforestation has not only contributed to the aesthetics but also has been serving as a 'sink' for the pollutants released from the station and thereby protecting the quality of ecology and environment in and around the projects. Thrust has also been given to bio-diesel plantation and around 4.8 lakh energy plants including Pongamia and Jatropha have already been planted. A pilot project for extraction of seeds from these bio-diesel plants has also been set up.

THE MARKET FOR ENVIRONMENTAL BIOTECHNOLOGY

The UK's Department of Trade and Industry estimated that 15–20 per cent of the global environmental market in 2001 was biotech-based, which amounted to about 250–300 billion US dollars and the industry is projected to grow by as much as ten-fold over the following five years. This expected growth is due to greater acceptance of biotechnology for clean manufacturing applications and energy production, together with increased landfill charges and legislative changes in waste management which also alter the UK financial base favourably with respect to bioremediation.

Biotechnology-based methods are seen as essential to help meet European Union (EU) targets for biowaste diversion from landfill and reductions in pollutants. Across the world the existing regulations on environmental pollution are predicted to be more rigorously enforced, with more stringent compliance standards implemented. All of this is expected to stimulate the sales of biotechnology-based environmental processing methods significantly and, in particular, the global market share is projected to grow faster than the general biotech sector trend, in part due to the anticipated large-scale EU aid for environmental clean-up in the new accession countries of Eastern Europe.

Other sources paint a broadly similar picture. The BioIndustry Association (BIA) survey, Industrial Markets for UK Biotechnology – Trends and Issues, published in 1999 does not quote any monetary sector values per year, but gives the size of the UK sector as employing 40 000 people in 1998 with an average yearly growth over 1995–98 of 20 per cent. Environmental biotech is reported as representing around 10 per cent of this sector. An Arthur Anderson report of 1997 gives the turnover of the UK biotech sector as 702 million pounds sterling in 1995/96, with a 50 per cent growth over three years. A 1998 Ernst and Young report on the European Life Sciences Sector says that the market for biotechnology products has the potential to reach 100 billion pounds sterling worldwide by 2005.

The Organisation for Economic Cooperation and Development (OECD) estimates that the global market for environmental biotechnology products and services alone will rise to some US$75 billion by the year 2000, accounting for some 15 to 25 per cent of the overall environmental technology market, which has a growth rate estimated at 5.5 per cent per annum. The UK potential market for environmental biotechnology products and services is estimated at between 1.65 and 2.75 billion US dollars and the growth of the sector stands at 25 per cent per annum, according to the Bio- Commerce Data European Biotechnology Handbook.

An unsourced quote found on a Korean University web site says that the world market size of biotechnology products and services was estimated to be approximately 390 billion US dollars in the year 2000. The benefits are not, however, confined to the balance sheet. The Organisation for Economic Cooperation and Development (OECD 2001) concluded that the industrial use of biotechnology commonly leads to increasingly environmentally harmonious processes and additionally results in lowered operating and/or capital costs. For years, industry has appeared locked into a seemingly unbreakable cycle of growth achieved at the cost of environmental damage.

The OECD investigation provides what is probably the first hard evidence to support the reality of biotechnology's long-heralded promise of alternative production methods, which are ecologically sound and economically efficient. A variety of industrial sectors including pharmaceuticals, chemicals, textiles, food and energy were examined, with a particular emphasis on biomass renewable resources, enzymes and bio-catalysis.

While such approaches may have to be used in tandem with other processes for maximum effectiveness, it seems that their use invariably leads to reduction in operating or capital costs, or both. Moreover, the research also concludes that it is clearly in the interests of governments of the developed and developing worlds alike to promote the use of biotechnology for the substantial reductions in resource and energy consumption, emissions, pollution and waste production it offers. The potential contribution to be made by the appropriate use of biotechnology to environmental and economic sustainability would seem to be clear.

The upshot of this is that few biotech companies in the environmental sector perceive problems for their own business development models, principally as a result of the wide range of businesses for which their services are applicable, the relatively low market penetration to date and the large potential for growth.

Competition within the sector is not seen as a major issue either, since the field is still largely open and unsaturated. Moreover, there has been a discernible tendency in recent years towards niche specificity, with companies operating in more specialised subarenas within the environmental biotechnology umbrella. Given the number and diversity of such possible slots, coupled with the fact that new opportunities and the technologies to capitalise on them, are developing apace, this trend seems likely to continue.

It is not without some irony that companies basing their commercial activities on biological organisms should themselves come to behave in such a Darwinian fashion. However, the picture is not entirely rosy. Typically the sector comprises a number of relatively small, specialist companies and the market is, as a consequence, inevitably fragmented. Often the complexities of individual projects make the application of 'standard' off-theshelf approaches very difficult, the upshot being that much of what is done must be significantly customised. While this, of course, is a strength and of great potential environmental benefit, it also has hard commercial implications which must be taken into account.

A sizeable proportion of companies active in this sphere, have no products or services which might reasonably be termed suitable for generalised use, though they may have enough expertise, experience or sufficiently perfected techniques to deal with a large number of possible scenarios. The fact remains that one of the major barriers to the wider uptake of biological approaches is the high perceived cost of these applications.

Part of the reason for this lies in historical experience. For many years, the solutions to all environmental problems were seen as expensive and for many, particularly those unfamiliar with the multiplicity of varied technologies available, this has remained the prevalent view. Generally, there is often a lack of financial resource allocation available for this kind of work and biotech providers have sometimes come under pressure to reduce the prices for their services as a result.

Greater awareness of the benefits of biotechnology, both as a means to boost existing markets and for the opening up of new ones, is an important area to be addressed. Many providers, particularly in the UK, have cited a lack of marketing expertise as one of the principal barriers to their exploitation of novel opportunities.

In addition, a lack of technical understanding of biotech approaches amongst target industries and, in some cases, downright scepticism regarding their efficacy, can also prove problematic. Good education, in the widest sense, of customers and potential users of biological solutions will be one major factor in any future upswing in the acceptance and utilisation of these technologies.

Modalities and Local Influences

Another of the key factors affecting the practical uptake of environmental biotechnology is the effect of local circumstances. Contextual sensitivity is almost certainly the single most important factor in technology selection and represents a major influence on the likely penetration of biotech processes into the marketplace.

Neither the nature of the biological system, nor of the application method itself, play anything like so relevant a role. This may seem somewhat unexpected at first sight, but the reasons for it are obvious on further inspection. While the character of both the specific organisms and the engineering remain essentially the same irrespective of location, external modalities of economics, legislation and custom vary on exactly this basis.

Accordingly, what may make abundant sense as a biotech intervention in one region or country, may be totally unsuited to use in another. In as much as it is impossible to discount the wider global economic aspects, disassociating political, fiscal and social conditions equally cannot be done, as the following example illustrates.

In 1994, the expense of bioremediating contaminated soil in the United Kingdom greatly exceeded the cost of removing it to landfill. Six years later, with successive changes of legislation and the imposition of a landfill tax, the situation has almost completely reversed. In those other countries where landfill has always been an expensive option, remediation has been embraced far more readily.

While environmental biotechnology must, inevitably, be viewed as contextually dependent, as the previous example shows, contexts can change. In the final analysis, it is often fiscal instruments, rather than the technologies, which provide the driving force and sometimes seemingly minor modifica-tions in apparently unrelated sectors can have major ramifications for the application of biotechnology.

The legal framework is another aspect of undeniable importance in this respect. Increasingly tough environmental law makes a significant contribution to the sector and changes in regulatory legislation are often enormously influential in boosting existing markets or creating new ones.

When legislation and economic pressure combine, as, for example, they have begun to do in the European Landfill Directive, the impetus towards a fundamental paradigm shift becomes overwhelming and the implications for relevant biological applications can be immense. There is a natural tendency to delineate, seeking to characterise technologies into particular categories or divisions. However, the essence of environmental biotechnology is such that there are many more similarities than differences. Though it is, of course, often helpful to view individual technology uses as distinct, particularly when considering treatment options for a given environmental problem, there are inevitably recurrent themes which feature throughout the whole topic. Moreover, this is a truly applied science. While the importance of the laboratory bench cannot be denied, the controlled world of research translates imperfectly into the harsh realities of commercial implementation. Thus, there can often be a dichotomy between theory and application and it is precisely this fertile ground which is explored in the present work.

In addition, the principal underlying approach of specifically environmental biotechnology, as distinct from other kinds, is the reliance on existing natural cycles, often directly and in an entirely unmodified form. Thus, this science stands on a foundation of fundamental biology and biochemistry. To understand the application, the biotechnologist must simply examine the essential elements of life, living systems and ecological circulation sequences. However engineered the approach, this fact remains true. In many respects, environmental biotechnology stands as the purest example of the newly emergent bioindustry, since it is the least refined, at least in terms of the basis of its action.

In essence, all of its applications simply encourage the natural propensity of the organisms involved, while seeking to enhance or accelerate their action. Hence, optimisation, rather than modification, is the typical route by which the particular desired end result, whatever it may be, is achieved and, consequently, a number of issues feature as common threads of individual technologies.

Integrated Approach

Integration is an important aspect for environmental biotechnology. One theme that will be developed throughout this book is the potential for different biological approaches to be combined within treatment trains, thereby producing an overall effect which would be impossible for any single technology alone to achieve. However, the wider goal of integration is not, of necessity, confined solely to the specific methods used. It applies equally to the underpinning knowledge that enables them to function in the first place and an understanding of this is central to the rationale behind this book. In some spheres, traditional biology has become rather unfashionable and the emphasis has shifted to more exciting sounding aspects of life science.

While the new-found concentration on 'Ecological processes', or whatever, sounds distinctly more 'Environ-mental', in many ways and somewhat paradoxically, it sometimes serves the needs of environmental biotechnology rather less well. The fundamentals of living systems are the stuff of this branch of science and, complex though the whole picture may be, at its simplest the environmental biotechnologist is principally concerned with a relatively small number of basic cycles. In this respect, a good working knowledge of biological processes like respiration, fermentation and photosynthesis, a grasp of the major cycles by which carbon, nitrogen and water are recycled and an appreciation of the flow of energy through the biosphere must be viewed as prerequisites.

The intent here has been neither to insult the readership by parading what is already well known, nor gloss over aspects which, if left unexplained, at least in reasonable detail, might only serve to confuse. However, this is expressly not designed to be a substitute for much more specific texts on these subjects, nor an entire alternative to a cohesive course on biology or biochemistry. The intention is to introduce and explain the necessary aspects and elements of various metabolic pathways, reactions and abilities as required to advance the reader's understanding of this particular branch of biotechnology.

A large part of the reasons for approaching the subject in this way is the fact that there really is no such thing as a 'typical' environmental biotechnologist. Practitioners come into the profession from a wide variety of disciplines and by many different routes. Thus, amongst their ranks are agronomists, biochemists, biologists, botanists, enzymologists, geneticists, microbiologists, molecular biologists, process engineers and protein technologists, all of whom bring their own particular skills, knowledge base and experiences. The applied nature of environmental biotechnology is obvious. While the science underlying the processes themselves may be as pure as any other, what distinguishes this branch of biological technology are the distinctly real-life purposes to which it is put. Hence, part of the intended function of this book is to attempt to elucidate the former in order to establish the basis of the latter.

At the same time, as any applied scientist will confirm, what happens in the field under operational conditions represents a distinct compromise between the theoretical and the practically achievable. At times, anything more than an approximation to the expected results may be counted as something of a triumph of environmental engineering.

Closing Remarks

The celebrated astronomer and biologist, Sir Fred Hoyle, said that the solutions to major unresolved problems should be sought by the exploration of radical hypotheses, while simultaneously adhering to well-tried and tested scientific tools and methods. This approach is particularly valid for environmental biotechnology. With new developments in treatment technologies appearing all the time, the list of what can be processed or remediated by biological means is ever changing. By the same token, the applications for which biotechnological solutions are sought are also subject to alteration. For the biotech sector to keep abreast of these new demands it may be necessary to examine some truly 'radical hypotheses' and possibly make use of organisms or their derivatives in ways previously unimagined. This is the basis of innovation; the inventiveness of an industry is often a good measure of its adaptability and commercial robustness.

CLASSIFICATION OF MICROBES

A microbe is any living thing that spends its life at a size visible sometimes only with a microscope. It is too tiny to be seen with the naked eye. Microbes are the oldest form of life on Earth. Some types have existed for billions of years. They may live as individuals or cluster together in communities. Microbes live in the water you drink, the food you eat and the air you breathe. Right now, billions of microbes are swimming in your belly and mouth and crawling on your skin! Don't worry, over 95 per cent of microbes are good for you. Microbes include bacteria, viruses, fungi, algae and protozoa. These single-cell organisms are invisible to the eye, but they can be seen with microscopes.

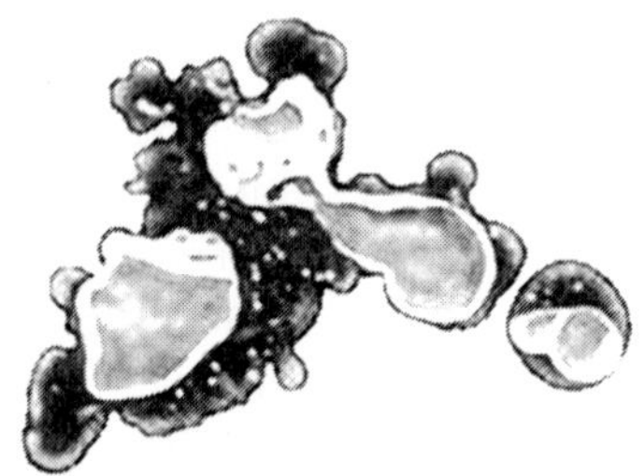

Fig. Microbe

The term microbe is short for microorganism, which means small organism. To help people understand the different types of microbes, they are grouped or classified in various ways. Microbes are very diverse and represent all the great kingdoms of life. In fact, in terms of numbers, most of the diversity of life on Earth is represented by microbes. The term microbe is short for microorganism, which means small organism. To help people understand the different types of microbes, they are grouped or classified in various ways. Microbes are very diverse and represent all the great kingdoms of life. In fact, in terms of numbers, most of the diversity of life on Earth is represented by microbes.

Here is an outline of the major groups of microorganisms:

- Viruses
- Bacteria
- Algae
- Fungi
- Protozoa

Viruses

A virus is too small to be seen without a microscope. A virus is basically a tiny bundle of genetic material carried in a shell called the viral coat. Some viruses have an additional layer around this coat called an envelope. That's basically all there is to viruses. There are thousands of different viruses that come in many shapes. Many are multi-sided or polyhedral. If you've ever looked closely at a cut gem, like the diamond in an engagement ring, you've seen an example of a polyhedral shape. Unlike the diamond in a ring, however, a virus does not taper to a point, but is shaped the same all around. Other viruses are shaped like spiky ovals or bricks with rounded corners. Some are like skinny sticks while others look like pieces of looped string. Some are more complex and shaped like little spaceship landing pods.

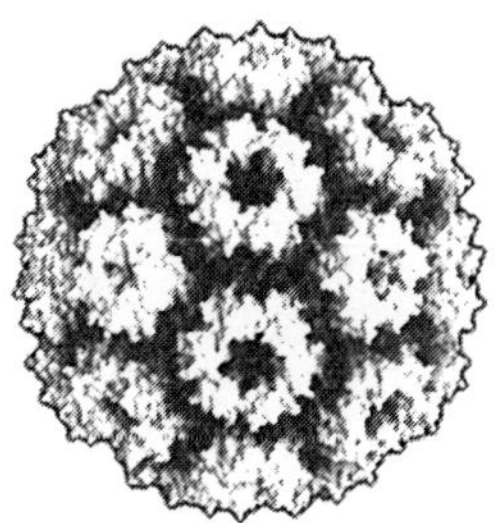

Fig. Viruses

Viruses are found on or in just about every material and environment on Earth from soil to water to air. They're basically found anywhere there are cells to infect. Viruses can infect every living thing. However, viruses tend to be somewhat picky about what type of cells they infect. Plant viruses are not equipped to infect animal cells, for example, though a certain plant virus could infect a number of related plants. Sometimes, a virus may infect one animal and do no harm, but cause a great deal of damage when it gets into a different but closely related animal.

Viruses exist to reproduce only. To do that, they have to take over suitable host cells. Upon landing on a suitable host cell, a virus gets its genes inside the cell either by tricking the host cell to pull it inside, or by connecting its viral coat with the host cell wall or membrane and releasing its genes inside, or by injecting their genes into the host cell's DNA. The viral genes are then copied many times, using the process the host cell would normally use to reproduce its own DNA.

The new viral genes then come together and assemble into whole new viruses. The new viruses are either released from the host cell without destroying the cell or eventually build up to a large enough number that they burst the host cell.

Bacteria

Bacteria consist of only one cell, but they're a very complex group of living things. Some bacteria can live in temperatures above the boiling point and in cold below the freezing point.

There are thousands of species of bacteria, but all of them are basically one of three different shapes. Some are rod- or stick-shaped; others are shaped like little balls. Others still are helical or spiral in shape. Some bacteria cells exist as individuals while others cluster together to form pairs, chains, squares or other groupings. Some bacteria can make their own food from sunlight, just like plants.

Also like plants, they give off oxygen. Other bacteria absorb food from the material they live on or in. Some of these bacteria can live off iron or sulfur! The bacteria that live in your stomach absorb nutrients from the digested food you've eaten.

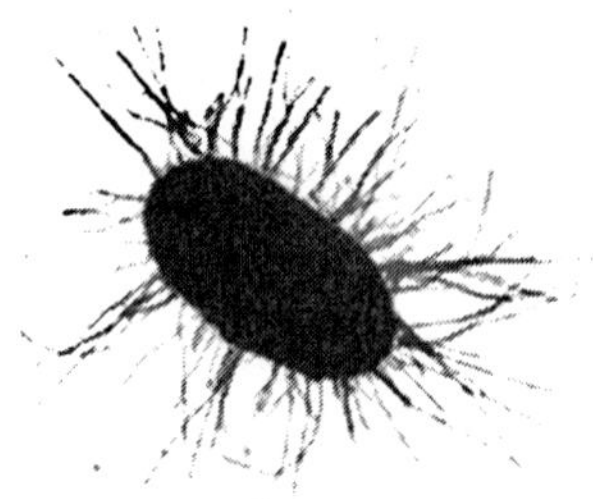

Fig. Bacteria

Some bacteria move about their environment by means of long, whip-like structures called flagella. They rotate their flagella like tiny outboard motors to propel themselves through liquid environments. They may also reverse the direction in which their flagella rotate so that they tumble about in one place. Other bacteria secrete a slime layer and ooze over surfaces like slugs. Others stay almost in the same spot.

Bacteria live on or in just about every material and environment on Earth from soil to water to air and from your body to the Arctic ice to the Sahara deserts. Each square centimeter of your skin averages about 100,000 bacteria. A single teaspoon of soil contains more than a billion (1,000,000,000) bacteria.

Algae

Algae are found in fresh and salt water around the world. They can also grow on rocks and trees and in soil when enough water is available.

Most algae are able to make energy from sunlight, like plants do. They produce a large amount of the oxygen we breathe. However, at some stages of their lives, some algae get their nutrients from other living things.

Diatoms are one kind of algae. They have hard shells made out of glass. When they die, these shells sink to the bottom of their watery environments. We mine deposits of these glass shells that formed hundreds of thousands of years ago to make abrasives, shiny road paint and grit in toothpaste. Diatoms come in all sorts of shapes—some are round and others are oval. Some look like leaves and others like fat commas.

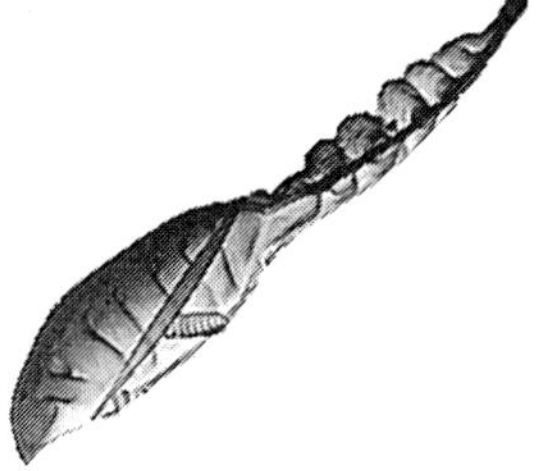

Fig. Algae

Because algae make so much oxygen, these microbes are very helpful. But sometimes certain kinds of algae can also grow in such large numbers that when they suddenly die off en masse, the breaking down of their cells by bacteria destroys the amount of dissolved oxygen in the water, hurting the animals and plants that live there.

Fungi

Fungi are organisms that scientists once confused with plants. However, scientists have found that, at the cell level, the fungi are more like animals than they are like plants. For one thing, fungi cannot synthesize their own food like plants do, but instead they eat other organisms as do animals.

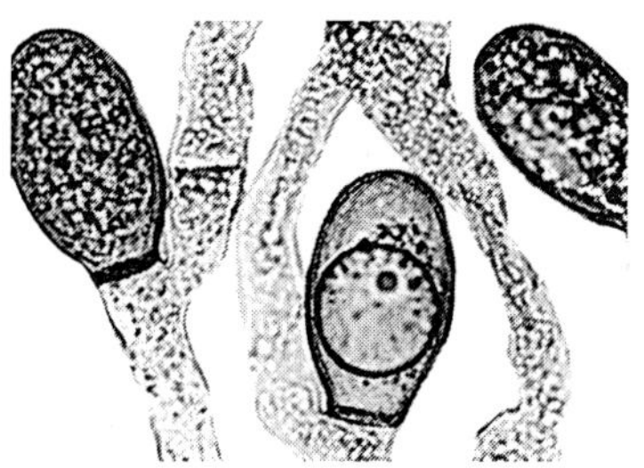

Fig. Fungi

Fungi come in a variety of shapes and sizes and different types. They can range from single cells to enormous chains of cells that can stretch for miles. Fungi include single-celled living things that exist individually, such as yeast and multicellular clusters, such as molds or mushrooms. Yeast cells look round or oval under a microscope. They're too small to see as individuals, but you can see large clusters of them as a white powdery coating on fruits and leaves.

Molds are described as filament-like because they form long filament-like, or thread-like, strands of cells called hyphae. These hyphae are what give mold colonies their fuzzy appearance. They also form the fleshy body, or mushroom, that some species grow. It may seem strange to think of something as big as a mushroom as a microbe. But the cells of the hyphae making up that mushroom arc connected in a closer way than the cells of other multicellular living things, like you and me, are.

Fungi usually grow best in environments that are slightly acidic. They can grow on substances with very low moisture. Fungi live in the soil and on your body, in your house and on plants and animals, in freshwater and seawater. A single teaspoon of topsoil contains about 120,000 fungi.

Fungi are basically stationary. But they can spread either by forming reproductive spores that are carried on wind and rain or by growing and extending their hyphae. Hyphae grow as new cells form at the tips, creating even longer chains of cells. Fungi absorb nutrients from living or dead organic matter that they grow on. They absorb simple, easily dissolved nutrients, such as sugars, through their cell walls. They give off special digestive enzymes to break down complex nutrients into simpler forms that they can absorb.

Some fungi are quite useful to us. We've used several kinds to make antibiotics to fight bacterial infections. These antibiotics are based on natural compounds the fungi produce to compete against bacteria for nutrients and space. We use baker's yeast, to make bread rise and to brew beer. Fungi break down dead plants and animals and keep the world tidier. We're exploring ways to use natural fungal enemies of insect pests to get rid of these bugs.

There are some dangerous fungi that cause diseases in plants, animals and people. Fungi ruin about a quarter to half of harvested fruits and vegetables each year.

CULTIVATION OF MICROBIAL

When microorganisms are cultivated in the laboratory, a growth environment called a medium is used. The medium may be purely chemical (a chemically defined medium), or it may contain organic materials, or it may consist of living organisms such as fertilized eggs.Microorganisms growing in or on such a medium form a culture. A culture is considered a pure culture if only one type of organism is present and a mixed culture if populations of different organisms are present. When first used, the culture medium should be sterile, meaning that no form of life is present before inoculation with the microorganism.

General microbial media. For the cultivation of bacteria, a commonly used medium is nutrient broth, a liquid containing proteins, salts and growth enhancers that will support many bacteria. To solidify the medium, an agent such as agar is added. Agar is a polysaccharide that adds no nutrients to a medium, but merely solidifies it. The medium that results is nutrient agar.

Many media for microorganisms are complex, reflecting the growth requirements of the microorganisms. For instance, most fungi require extra carbohydrate and an acidic environment for optimal growth. The medium employed for these organisms is potato dextrose agar, also known as Sabouraud dextrose agar. For protozoa, liquid media are generally required and for rickettsiae and viruses, living tissue cells must be provided for best cultivation.

For anaerobic microorganisms, the atmosphere must be oxygen free. To eliminate the oxygen, the culture media can be placed within containers where carbon dioxide and hydrogen gas are generated and oxygen is removed from the atmosphere. Commercially available products achieve these conditions. Anaerobic chambers can also be used within closed compartments and technicians can manipulate culture media within these chambers. To encourage carbon dioxide formation, a candle can be burned to use up oxygen and replace it with carbon dioxide.

Special microbial media. Certain microorganisms are cultivated in selective media. These media retard the growth of unwanted organisms while encouraging the growth of the organisms desired. For example, mannitol salt agar is selective for staphylococci because most other bacteria cannot grow in its high-salt environment. Another selective medium is brilliant green agar, a medium that inhibits Gram-positive bacteria while permitting Gram-negative organisms such as *Salmonella* species to grow.

Still other culture media are differential media. These media provide environments in which different bacteria can be distinguished from one another. For instance, violet red bile agar is used to distinguish coliform bacteria such as *Escherichia coli* from non-coliform organisms. The coliform bacteria appear as bright pink colonies in this media, while non-coliforms appear a light pink or clear. Certain media are both selective and differential.

For instance, MacConkey agar differentiates lactose-fermenting bacteria from non-lactose-fermenting bacteria while inhibiting the growth of Gram-positive bacteria. Since lactose-fermenting bacteria are often involved in water pollution, they can be distinguished by adding samples of water to MacConkey agar and waiting for growth to appear.

In some cases, it is necessary to formulate an enriched medium. Such a medium provides specific nutrients that encourage selected species of microorganisms to flourish in a mixed sample. When attempting to isolate *Salmonella* species from fecal samples, for instance, it is helpful to place a sample of the material in an enriched medium to encourage *Salmonella* species to multiply before the isolation techniques begin.

In order to work with microorganisms in the laboratory, it is desirable to obtain them in pure cultures. Pure cultures of bacteria can be obtained by spreading bacteria out and permitting the individual cells to form masses of growth called colonies. One can then pick a sample from the colony and be assured that it contains only one kind of bacteria. Cultivating these bacteria on a separate medium will yield a pure culture. To preserve microbial cultures, they may be placed in the refrigerator to slow down the metabolism taking place. Two other methods are deep-freezing and freeze-drying.

For deep-freezing, the microorganisms are placed in a liquid and frozen quickly at temperatures below –50°C. Freeze-drying (lyophilization) is performed in an apparatus that uses a vacuum to draw water off after the microbial suspension has been frozen. The culture resembles a powder and the microorganisms can be preserved for long periods in this condition.

STERILIZATION

Wet Heat

Wet heat is the most dependable procedure for the destruction of all forms of microbial life. Steam sterilization generally denotes heating in an autoclave employing saturated steam under a pressure of approximately 15 psi to achieve a chamber temperature of at least 121°C (250°F). The critical factors in insuring the reliability of this sterilization method is:

- Proper temperature and time; and
- The complete replacement of the air with steam (*i.e.* no entrapment of air).

Some autoclaves utilize a steam activated exhaust valve that remains open during the replacement of air by live steam until the steam triggers the valve to close. Others utilize a pre-cycle vacuum to remove air prior to steam introduction.

Physical controls such as pressure gauges and thermometers are widely used but are considered secondary methods of insuring sterilization. The use of appropriate biological indicators at locations throughout the autoclave is considered the best indicator of sterilization. The biological indicator most widely used for wet heat sterilization is Bacillus stearothermophilus spores.

Dry Heat

Dry heat is less efficient than wet heat sterilization and requires longer times and/or higher temperatures. The specific times and temperatures must be determined for each type of material being sterilized. Generous safety factors are usually added to allow for the variables that can influence the efficiency of this method of sterilization. The moisture of the sterilization environment as well as the moisture history of organisms prior to heat exposure appear to affect the efficiency of dry heat sterilization.

Higher temperatures and shorter times may be used for heat resistant materials. The heat transfer properties and the spatial relation or arrangement of articles in the load are critical in insuring effective sterilization.

The advantage of wet heat is a better heat transfer to and into the cell resulting in overall shorter exposure time and lower temperature. Steam sterilization uses pressurized steam at 121-132° C (250-270° F) for 30 or 40 minutes. This type of heat kills all microbial cells including spores, which are normally heat resistant. In order to accomplish the same effect with dry heat in an oven, the temperature needs to be increased to 160-170° C (320-338° F) for periods of 2 to 4 hours.

General Procedures

All materials, equipment, apparatus contaminated with or containing potentially hazardous organisms should be steam sterilized before being washed and stored, or discarded. Autoclaving is the preferred method. Each individual working with biohazardous materials is responsible for sterilization of materials before disposal.

Biohazardous materials should not be placed in autoclaves overnight in anticipation of autoclaving the next day. To minimize hazard to firemen or disaster crews, all biohazardous materials must be placed in an appropriately marked refrigerator or incubator, sterilized, or otherwise confined at the close of each work day.

All autoclaves must be certified for operating efficiency by the periodic use of biological indicator controls and records maintained for three years. Special precautions should be taken to prevent accidental removal of material from an autoclave before it has been sterilized or the simultaneous opening of both doors on a double door autoclave. Dry hypochlorites, or any other strong oxidizing material, must not be autoclaved with organic materials such as a paper, cloth, or oil:

Oxidizer + Organic material + Heat = Possible Explosion.

All floors, laboratory benches and other surfaces in buildings where biohazardous materials are handled should be disinfected as often as deemed necessary by the supervisor.

The surroundings should be disinfected after completion of operations involving plating, pipetting, centrifuging and similar procedures with biohazardous materials. It is the responsibility of the supervisor to determine that the disinfectant and the time and method of exposure is effective against the biological agent(s) used in the facility.

Floor drains should be flooded with water or disinfectant at least once each week in order to fill traps and thus prevent the back flow of sewer gases. Floor cleaning procedures which minimize the generation of aerosols should be used. Wet mopping or wet vacuum pick up is recommended.

Water used to mop floors should contain a disinfectant or disinfectant-detergent. (Dry mopping or dusting should be avoided). Where wet procedures are not practical, dry vacuum cleaning with a HEPA filter on the exhaust, sweeping compound used with push brooms, or dry dust mop heads treated to suppress aerosolization may be used. Stock solutions of suitable disinfectants will be maintained in each laboratory for disinfection purposes. NOTE: working dilutions of certain disinfectants have short shelf lives and must be prepared on a regular basis.

Sterilization Procedures

General criteria for sterilization of typical materials are presented below. Supervisors are encouraged to review the type of materials being handled and to establish standard conditions for sterilization. Treatment conditions to achieve sterility will vary in relation to the volume of material treated, volume of the autoclave, the contamination level, the moisture content and other factors.

Steam Autoclave

- *Laundry:* 250°F (121°C) for a minimum of 30 min.
- *Trash:* 250°F (121°C) for at least 45 minutes per bag. Size of the autoclave and of the bags greatly effect sterilization time. Large bags in a small autoclave may require 90 minutes or more.
- *Glassware:* 250°F (121°C) for a minimum of 25 min.
- *Liquids:* 250°F (121°C) for 25 minutes for each gallon.
- *Animals and bedding:* Steam autoclaving not recommended (sterilization time required would be at least 8 hours). Incineration in an approved facility is the recommended treatment of these wastes.

MICROSCOPES

Since microorganisms are invisible to the unaided eye, the essential tool in microbiology is the microscope. One of the first to use a microscope to observe microorganisms was Robert Hooke, the English biologist who observed algae and fungi in the 1660s. In the 1670s, Anton van Leeuwenhoek, a Dutch merchant, constructed a number of simple microscopes and observed details of numerous forms of protozoa, fungi and bacteria.

During the 1700s, microscopes were used to further elaborate on the microbial world and by the late 1800s, the sophisticated light microscopes had been developed. The electron microscope was developed in the 1940s, thus making the viruses and the smallest bacteria (for example, rickettsiae and chlamydiae) visible. Microscopes permit extremely small objects to be seen, objects measured in the metric system in micrometers and nanometers. A micrometer (ìm) is equivalent to a millionth of a meter, while a nanometer (nm) is a billionth of a meter. Bacteria, fungi, protozoa and unicellular algae are normally measured in micrometers, while viruses are commonly measured in nanometers. A typical bacterium such as *Escherichia coli* measures about two micrometers in length and about one micrometer in width.

Optical Microscopy

Optical or light microscopy involves passing visible light transmitted through or reflected from the sample through a single or multiple lenses to allow a magnified view of the sample.

The resulting image can be detected directly by the eye, imaged on a photographic plate or captured digitally. The single lens with its attachments, or the system of lenses and imaging equipment, along with the appropriate lighting equipment, sample stage and support, makes up the basic light microscope.

Limitations of Optical Microscopy

Limitations of standard optical microscopy (bright field microscopy) lie in three areas;

- The technique can only image dark or strongly refracting objects effectively.
- Diffraction limits resolution to approximately 0.2 micrometre.
- Out of focus light from points outside the focal plane reduces image clarity.

Live cells in particular generally lack sufficient contrast to be studied successfully, internal structures of the cell are colourless and transparent. The most common way to increase contrast is to stain the different structures with selective dyes, but this involves killing and fixing the sample. Staining may also introduce artifacts, apparent structural details that are caused by the processing of the specimen and are thus not a legitimate feature of the specimen.

These limitations have, to some extent, all been overcome by specific microscopy techniques which can non--invasively increase the contrast of the image. In general, these techniques make use of differences in the refractive index of cell structures. It is comparable to looking through a glass window: you (bright field microscopy) don't see the glass but merely the dirt on the glass.

There is however a difference as glass is a more dense material and this creates a difference in phase of the light passing through. The human eye is not sensitive to this difference in phase but clever optical solutions have been thought out to change this difference in phase into a difference in amplitude (light intensity).

OPTICAL MICROSCOPY TECHNIQUES

Bright Field Optical Microscopy

Bright field microscopy is the simplest of all the light microscopy techniques. Sample illumination is via transmitted white light, *i.e.* illuminated from below and observed from above. Limitations include low contrast of most biological samples and low apparent resolution due to the blur of out of focus material. The simplicity of the technique and the minimal sample preparation required are significant advantages.

Oblique Illumination and what it Means

The use of oblique (from the side) illumination gives the image a 3-dimensional appearance and can highlight otherwise invisible features. A more recent technique based on this method is Hoffmann's modulation contrast, a system found on inverted microscopes for use in cell culture. Oblique illumination suffers from the same limitations as bright field microscopy (low contrast of many biological samples; low apparent resolution due to out of focus objects), but may highlight otherwise invisible structures.

Dark field Optical Microscopy

Dark field microscopy is a technique for improving the contrast of unstained, transparent specimens. Darkfield illumination uses a carefully aligned light source to minimise the quantity of directly-transmitted (un-scattered) light entering the image plane, collecting only the light scattered by the sample. Darkfield can dramatically improve image contrast—especially of transparent objects—while requiring little equipment setup or sample preparation. However, the technique does suffer from low light intensity in final image of many biological samples and continues to be affected by low apparent resolution.

Rheinberg illumination is a special variant of dark field illumination in which transparent, coloured filters are inserted just before the condenser so that light rays at high aperture are differently coloured than those at low aperture (*i.e.* the background to the specimen may be blue while the object appears self-luminous yellow). Other colour combinations are possible but their effectiveness is quite variable.

Phase Contrast Optical Microscopy

More sophisticated techniques will show differences in optical density in proportion. Phase contrast is a widely used technique that shows differences in refractive index as difference in contrast. It was developed by the Dutch physicist Frits Zernike in the 1930s The nucleus in a cell for example will show up darkly against the surrounding cytoplasm. Contrast is excellent; however it is not for use with thick objects. Frequently, a halo is formed even around small objects, which obscures detail.

The system consists of a circular annulus in the condenser which produces a cone of light. This cone is superimposed on a similar sized ring within the phase-objective. Every objective has a different size ring, so for every objective another condenser setting has to be chosen. The ring in the objective has special optical properties: it first of all reduces the direct light in intensity, but more importantly, it creates an artificial phase difference of about a quarter wavelength. As the physical properties of this direct light have changed, interference with the diffracted light occurs, resulting in the phase contrast image.

Differential Interference Contrast Microscopy

Superior and much more expensive is the use of interference contrast. Differences in optical density will show up as differences in relief. A nucleus within a cell will actually show up as a globule in the most often used differential interference contrast system according to Georges Nomarski. However, it has to be kept in mind that this is an optical effect and the relief does not necessarily resemble the true shape! Contrast is very good and the condenser aperture can be used fully open, thereby reducing the depth of field and maximizing resolution. The system consists of a special prism in the condenser that splits light in an ordinary and an extraordinary beam. The spatial difference between the two beams is minimal (less than the maximum resolution of the objective). After passage through the specimen, the beams are reunited by a similar prism in the objective.

In a homogeneous specimen, there is no difference between the two beams and no contrast is being generated. However, near a refractive boundary (say a nucleus within the cytoplasm), the difference between the ordinary and the extraordinary beam will generate a relief in the image. Differential interference contrast requires a polarized light source to function; two polarizing filters have to be fitted in the light path, one below the condenser (the polarizer) and the other above the objective (the analyzer).

Note: In cases where the optical design of a microscope produces an appreciable lateral separation of the two beams we have the case of classical interference microscopy, which does not result in relief images, but can nevertheless be used for the quantitative determination of mass-thicknesses of microscopic objects.

Environmental Issues

There is concern that halons are being broken down in the atmosphere to bromine, which reacts with ozone, leading to depletion of the ozone layer (this is similar to the case of chlorofluorocarbons such as freon). These issues are complicated: the kinds of fires that require halon extinguishers to be put out will typically cause more damage to the ozone layer than the halon itself, not to mention human and property damage. However, fire extinguisher systems must be tested regularly, and these tests may lead to damage. As a result, some regulatory measures have been taken, and halons are being phased out in most of the world.

History

Carbon tetrachloride was used in fire extinguishers and glass (anti)-"fire grenades"from the latenineteenth century until around the end of World War II. Experimentation with halogenated hydrocarbons for fire suppression on military aircraft began at least as early as the 1920s. During World War II, various early halons were in standard use in aircraft by some combatants, but these early halons suffered from excessive toxicity. Nevertheless after the war they slowly became more common in civil aviation as well.

In the 1960s, fluorinated halons became available, and were quickly recognised as one of the most effective fire fighting materials discovered. Much early research with Halon 1301 was conducted under the auspices of the US

Armed Forces, while Halon 1211 was initially mainly developed in the UK. By the late 1960s, they were standard in many applications where water and dry powder extinguishers posed a threat of damage to the protected property, including computer rooms, telecommunications switches, Labouratories, museums and art collections. Beginning with warships in the 1970s, halons also progressively came to be associated with rapid knockdown of severe fires in confined spaces with minimal risk to personnel.

By the early 1980s, halons were in common use on aircraft, ships and large vehicles, as well as in computer facilities and galleries. However, concern was beginning to be felt about the possible impact of chlorinated and brominated hydrocarbons on the ozone layer. The Vienna Convention on Ozone Layer Protection did not cover halons, as it was felt that emergency discharge of systems was too small in volume to produce a significant impact, and too important to human safety for restriction. However, by the time of the Montreal Protocol it was realised that discharges during system tests and maintenance accounted for substantially larger volumes than emergency discharges, and so halons were brought into the treaty, but with many exceptions.

Halons have been generally phased out and the possession of halon equipment is prohibited in some countries like the Netherlands and Belgium from January 1, 2004, based on the Montreal Protocol and guidelines of the European Union. Production of new stocks has ceased in most (probably all) countries as of 1994. However many countries still require aircraft to be fitted with halon fire suppression systems, as no safe and completely satisfactory alternative has been discovered for this application. There are also a few other highly specialised users.

Bibliography

A G Brecket: *A Handbook on Nanotechnology*, Dominant Publication, Delhi, 2008

A. S. Edelstein and R. C. Cammarata: *Nanomaterials: Synthesis, Properties, and Applications,* Philadelphia, Institute of Physics Publication, Bristol, 1998.

Abdullahi, A.: *Statistical Mechanical Modeling and Its Application to Nanosystems*, Philadelphia, PA: University of Pennsylvania Press, 1998.

Ach J.S. and Siep L.: *Nano-bio-ethics: Ethical Dimensions of Nanobiotechnology,* Transaction Publishers London, 2006.

Aggarwal, S.K.: *A Textbook of Optics*, Wisdom Press, Delhi, 2012.

Ahmad, Lone: *Nanoscience: Nanotechnologies and Nanophysics,* South Kashmir Press, Jammu Kashmir, 2006.

Ajay Kakkar: *Know All About Nanotechnology,* Abhishek Publications, Delhi, 2010.

Allen, Robin: *An Introduction to Micro and Nanotechnology*, Oxford University Press, New York, 2002.

Allhoff F. and Lin P.: *Nanotechnology and Society: Current and Emerging Ethical Issues,* Springer The Netherlands, 2009.

Anand Prakash Singh: *Nanotechnology : A Broad Spectrum,* Axis Publications, Delhi, 2010.

Balakrishna Rao and K. Krishna Reddy: *Nanotechnology in Health and Life Sciences,* Campus Books International, Delhi, 2007.

Baldev Raj, B.S. Murty, P. Shankar, B.B. Rath and James Murday: *Textbook of Nanoscience and Nanotechnology,* Orient Blackswan, Delhi, 2011.

Balzani, Vincenzo, M. Venturi, and A. Credi: *Molecular devices and machines: a journey into the nano world,* Wiley-VCH, Weinheim, 2003.

Barker, J.R., S. Roy, S. Babiker, and A. Asenov: *Proceedings of the International Conference on Quantum Devices and Circuits,* Singapore, World Scientific, 1997.

Basnet, J.: *Unique Features and Properties of Nanostructured Materials,* Indian J Community Med, Delhi, 2008.

Bhanu, T.K. Chandra and Vasudev Bhatnagar: *Nanostructure Science and Technology*, Campus Books, Delhi, 2009.

Bharadwaj, Ashish : *The Nanoscope: Encyclopedia of Nanoscience and Nanotechnology*, Pentagon Press, Delhi, 2005.

Bhatia, A.S. : *Nanowires and Nanobelts*, Deep and Deep, Delhi, 2011.

Bhatnagar, Vasudev: *Nanostructure Science and Technology*, Campus Books, Delhi, 2009.

Brune: *Nanotechnology: Assessment and Perspectives*, Springer, Delhi, 2010.

Bueno, O.: *The Drexler-Smalley Debated on Nanotechnology: Incommensurability at Work*, Willey, Delhi, 2004.

Cameron N. M. de S. and Mitchell M. E.: *Nanoscale: Issues and Perspectives for the Nano Century,* John Wiley & Sons Hoboken, 2007.

Carl C. Koch: *Nanostructured Materials: Processing, Properties, and Potential Applications*, Noyes Publications/William Andrew Pub., Norwich, 2002.

Carlsson, S.B., T. Junno, H. Xu, L. Montelius: *Extended abstracts of the 3rd International Workshop on Quantum Functional Devices*, Tokyo, R&D Association for FED, 1997.

Cerofolini, G.F.: *Nanoscale Devices,* Springer-Verlag, Berlin, 2009.

Charles Poole and Frank Owens: *Introduction To Nanotechnology,* Wiley, Delhi, 2010.

Choi, Hee Cheul: *Studies of Porous Semiconducting Materials and their Applications in Nanotechnology*, University Microfilms, 2002.

Cleland, Andrew N.: *Foundations of Nanomechanics: From Solid-State Theory to Device Applications*, Springer, New York, 2003.

Coryton J.: *The Law of Letters-patent,* H. Sweet London, 1855.

Index